Wine Talk

Wine Talk

An Enthusiast's Take on the People, the Places, the Grapes, and the Styles

Raymond Blake

Skyhorse Publishing

Skyhorse Publishing books may be purchased in bulk at special discounts for sales promotion, corporate gifts, fund-raising, or educational purposes. Special editions can also be created to specifications. For details, contact the Special Sales Department, Skyhorse Publishing, 307 West 36th Street, 11th Floor, New York, NY 10018 or info@ skyhorsepublishing.com.

Skyhorse® and Skyhorse Publishing® are registered trademarks of Skyhorse Publishing, Inc.®, a Delaware corporation.

Visit our website at www.skyhorsepublishing.com.

10 9 8 7 6 5 4 3 2 1

Library of Congress Cataloging-in-Publication Data is available on file.

Cover design by David Ter-Avanesyan
Cover image by Getty Images

ISBN: 978-1-5107-6702-7
Ebook ISBN: 978-1-5107-6751-5

Printed in the United States of America

For Fionnuala

CONTENTS

ACKNOWLEDGMENTS   ix

INTRODUCTION   xi

CHAPTER 1 IN THE BEGINNING   1

CHAPTER 2 VINE TO GLASS   11

CHAPTER 3 BUBBLES   27

CHAPTER 4 LEGACY WINES   47

CHAPTER 5 ACROSS THE POND   63

CHAPTER 6 DOWN UNDER   79

CHAPTER 7 WHEN MONSTERS ROAMED   99

CHAPTER 8 ON LIGHTER FEET   115

CHAPTER 9 ACCESSORIZE   129

CHAPTER 10 AND TO EAT?   145

CHAPTER 11 GOLDEN WONDERS   161

CHAPTER 12 WHAT NEXT?   175

INDEX   193

ACKNOWLEDGMENTS

This book is slightly unusual in that it didn't rely on the usual roster of travel, winery visits, interviews, tastings, and meetings for its material. That was all collected and collated previously, ready to provide the bedrock for the writing. A greater resource, however, was the generosity of friends and colleagues.

This supplied the essential element, without which this book could not have been written. The host of treasured bottles that generous friends shared over the years provided, in addition to great enjoyment, an indispensable reference point, a calibration of palate and opinion that granted conviction to the words that follow. I count myself extremely fortunate in this regard because now that many, though not all, of the world's best wines are priced well beyond budget, this was my only hope of experiencing them. I could not have done without this passive assistance, for it allowed me to experience the world's greatest bottles, giving a benchmark without which this book would have been much the lesser.

Included in their number are: Pat Blake, Jean-Claude Bernard, Olivier Bernard, the Boisset Family, David Browne, Stephen Carrier, John Carroll, Prof John Gaskin, John Glen, Juris Grinbergs, Phillip Jones, Michael and Kate Hayes, Bill Kelly, Anthony and Olive Hamilton Russell, Ivan Healy, Dougie Heather, Paul Hunt, Andrew Keaveney, Noel Kierans, Des Lamph, Peter and Margaret Lehmann, Alf Mauff, Andrew and Sue Mead, Donal Morris, Monica Murphy, Martin Naughton, Micheál O'Connell, Eddie O'Connor, James O'Connor, Patrick O'Connor, Mark O'Mahony, Eugene and Lynette O'Sullivan, Jim Pilkington, Randall and Carol Plunkett, Joël Provence and Patricia Bouchey, Lochlann and Brenda Quinn, Jo Rodin, the Rowand Family, the Seysses Family, Brian Shiels, Dómhnal Slattery, Brian Spelman, Steven Spurrier, Jim Tunney, Dr Mahendra Varma, Aubert and Pamela de Villaine, the Vincent Family, David Whelehan, and Michael Yon.

With apologies to any I may have inadvertently overlooked. A good glass awaits anyone so omitted.

I am also indebted to my parents Gay and Frank, and sisters Barbara and Margaret, who never failed to provide an enquiring, encouraging word when needed.

A special word of thanks to Sharon Bowers and Julie Ganz. Sharon's advice was always astute and to the point, while Julie and the team at Skyhorse did a fine job of turning an unwieldy Word document into this neat book.

Which is not to forget the inestimable contribution made by my wife, Fionnuala, who also shared many treasured bottles, bought years before we met. When my belief faded to zero, hers never wavered. Wise counsel and silent understanding, combined with endless encouragement, kept the writing moving forward when it threatened to stall. As with my previous wine books, this one would have been impossible without her.

I must finish with a posthumous word of thanks to my late, great friend and colleague, Tomás Clancy. If ever I wanted my thoughts clarified and given focus, enthusiastically validated, I only had to run them past Tomás. When I outlined the scope of this book to him he immediately burst forth with a torrent of encouraging advice, advice that sustained me right to the end. He was a bottomless well of positivity. The deadline for this book was the first anniversary of his death. I hope it stands as a modest tribute of gratitude to him.

INTRODUCTION

"Is Pinot Noir a place?" the interviewer asked. "No, it's a grape, the grape of red Burgundy," I answered. "And Chardonnay, is it a place?" she continued. "As it happens, yes, though it's best known as a grape, the grape of white Burgundy." I further explained that, at their best, they produce incomparably brilliant still wines—and that in Champagne, Pinot Noir and Chardonnay go together to produce the world's greatest sparkling wines. This prompted disbelief and a request to explain. So I did—by way of tale and anecdote, to make the message sing.

Wine Talk does not aim to "take the mystery out of wine," a line championed in too much wine writing. Take the mystery out of wine? Why? That's the fun part. No mystery, no fun. Wine without mystery has a name—water. Wine should not be demystified to encourage newcomers—they should be told it is a subject worthy of endless investigation and debate. That's what makes wine fascinating and keeps us coming back again and again. The certain knowledge we will never know or understand it all is what sustains and nourishes interest, what gives wine enduring attraction. Take the mystery out of wine? Never!

This book is for that interviewer and the many millions of people who love a glass of wine and would like to know more without engaging in formal study. It is for those who drink their wine without ceremony but with some interest. For those who have been put off by highfalutin terminology and forbidding ritual. For those who want the message simplified but not dumbed down.

The material is culled from twenty-five years globetrotting the world's vineyards, always keen to see what was around the next corner, always keen to discover the next good bottle. Wine is, or can be, more than a beverage, more than a means of ingesting alcohol. It tells a story—of the people who made it and the place it comes from; and the message it carries sits mute in the bottle, full of promise and potential, until we fire the starting gun by pulling the cork. People and place are the markers that give it distinction and uniqueness, wine with character.

Wine also captures a moment in time and carries the stamp of its birth year like DNA. Additionally, each bottle of the same wine will tell a marginally different story because it is drunk at a different time in its evolution and in a different context. Its message changes subtly all the time. It loses everything, however, if it is simply an industrial muddle of anonymous flavors that could have been made by anyone, anywhere. Then, it is just another beverage and no special pleading can be made for it.

This book elaborates and interprets wine's story by way of broad brushstroke enthusiasms rather than expert diktats. It is an opinion-driven reflection on the world of wine that steers a course between the po-faced intensity with which some wine wisdom is dispensed and the hokum and hearsay that surrounds any discussion with alcohol at its heart. *Wine Talk* is not a textbook. It is written from the standpoint of an enthusiastic consumer who has made a career communicating its joys.

Notwithstanding all that has been said and written about wine in recent years, there is still a level of mild perplexity, perhaps even confusion, among passively interested consumers. The actively interested are well catered to. They have done the courses, read the books, and visited the regions. I want to inform and entertain the others in an easy-to-assimilate fashion. Tell them a story, paint pen portraits. They don't read formal wine articles, yet they are interested and want to know enough to banish a feeling of panic when tasked to make a choice from a wine list in a restaurant or a wall of wine in a retail outlet.

Many are intimidated by the rituals and beliefs that still surround wine. The old shibboleths die hard: that all wine improves with age; that red wine and cheese is a match made in heaven; that an open bottle will spoil overnight; that only maiden aunts drink Sherry and cussed colonels Port; that Champagne isn't really wine and is for celebration only; that new world wines are all very fine but French is better by definition.

Some ceremony enhances enjoyment—correct temperature, good glassware, appropriate food matches, and so forth—but there is only one written-in-stone rule to guarantee full enjoyment from your wine, so obvious that it is seldom mentioned. Wine must be shared with like-minded individuals. You can get everything else right but if the company, and hence the mood and context, is wrong, the wine will sour in the mouth. This is the most crucial requirement. To repeat, wine is for sharing and choice of companions is of prime importance; nothing else matters if this is not right. Hobbling ourselves with strictures about temperature and glasses only serves to blind us to this timeless truth. We must not get too precious about wine, but we must get the company right.

Context for me has always been paramount, and I am hoping this book puts wine in context. The aim is to trigger enthusiasm, not endow expertise, to foster curiosity and a desire to know more, to enhance people's engagement with wine. I revel in wine, telling its tale, delighting in its restorative powers and celebrating its inspiring effects. If this book fires enthusiasm in the reader, then it will be a success. Read it with a glass in hand. Revel in wine.

1

IN THE BEGINNING

In the beginning, there was Rosemount Chardonnay, Château Chasse-Spleen, and Warre's Vintage Port—a case of the first, a case of the second, and three of the third, all bought around the mid-1980s. Prior to that I only bought wine by the bottle, so this was a big step forward. The Rosemount was bottled sunshine and lasted a few months. The Chasse-Spleen was the excellent 1982 and lasted a decade. The Warre's was 1977, purchased to celebrate the turn of the millennium but that was a damp squib, so some bottles remain. As a "starter pack," it could hardly have been bettered. The trio covered many bases: the Rosemount was exuberant (drinkable in T-shirt and shorts), the Chasse-Spleen was reserved (it might need a shirt and tie), and the Warre's was commanding (best to wear a tux).

Though Australian wine was shipped to Europe in the early 1900s, it was in the 1980s that it took off, surging onto the public radar on a wave of exuberant fruit that challenged: you-can't-not-like-me. Old-timers, their palates corroded by decades of exposure to meager wines, struggled not to be charmed, defaulting to the objection that the wines wouldn't age, a grumble that missed the point. These wines cried out to be drunk and, short of hanging a "Drink me quick" label around the neck, that is what the bright and breezy packaging encouraged.

I lapped up the Rosemount, reveling in the flavorful swirl of succulent tropical fruit, shot through with vanilla, and finishing slightly lush. The texture was polished, with no hard edges—"fruit juice as wine" some sneered—but I didn't listen, it was impossible to resist. There weren't hidden nuances or complex depths; it *was* bottled sunshine, and every sip suggested another. The dozen bottles were drunk quickly, all resolve to eke them out abandoned; it simply charmed you, and wine wasn't meant to do that. (Also it aged better than expected, as proved by a bottle of the 1989 drunk in 2001.)

The charm started before the bottle was opened, courtesy of the cleverest label ever to adorn a wine bottle. Square, but turned through forty-five degrees to render it diamond-shaped, it was canary yellow and could be recognized from the far end of a supermarket aisle. It was a classy trick compared to some of the zany labels concocted today to catch the consumer eye. That yellow label signaled delight before a drop was drunk, and it didn't make a promise the wine couldn't deliver. They walked hand in hand. Since then the Rosemount name has lost luster, thanks to a series of ownership changes, with the diamond label flogged and cheapened by ubiquity. The excitement is gone, but for a while Rosemount shone like a shooting star.

Rosemount and Chasse-Spleen were opposites in every respect, from color to grape to style to origin. One was the brash upstart, the other a respected member of wine's gentry. Bordeaux insiders knew Chasse-Spleen as a sure thing that regularly punched above its attractive price—and the 1982 vintage did not disappoint. It would never be described as bottled sunshine, yet neither was it a coarse wine, examples of which were abundant in the 1980s. The fruit was properly ripe, there was reserve and elegance, and no trumpeting of its charms. To boot, I loved the fact, or perhaps the fiction, that Lord Byron named it for its ability to chase away the spleen or ill humors, leaving the drinker untroubled by life's woes.

The Chasse-Spleen was quiet but not dumb. It spoke clearly, and its signature was a soft savor that lay somewhere between fruit and earth. Repeated sipping revealed subtle variations, with its true class showing in the satisfying aftertaste that left the palate replete yet fresh. As the years passed, the wine evolved gently, the vigor subsided, and the components folded into one another until it became difficult to tell them apart. Tannin, acidity, and fruit melded to reveal a new "sweetness" that added further attraction.

Australia dazzled the wine world in the 1980s while Bordeaux emerged from a century in the doldrums. It is difficult today, after four decades of bounding prosperity, to grasp the dismal days that Bordeaux had endured. A litany of tribulation started with the vine diseases, oidium and phylloxera, then came world war, Prohibition in the United States, economic depression, more war, and a frost that destroyed vast swathes of vineyard in 1956, finishing with the economic turmoil caused by the oil crisis of the 1970s. The châteaux still presented grand facades to the world but behind the fairy-tale ornamentation the working facilities were in need of investment and renovation.

The city mirrored the neglect—it was shoddy, its splendid architecture careworn and dowdy. History lay heavy on Bordeaux, a circumstance unknown to the jaunty new world upstarts. They weren't separated only by geography but by winemaking philosophy too. One represented the new thrusting face of wine and the other was a relic resting on old glories. One swaggered, one staggered. Bordeaux was capable of sublime levels of quality when on song but too often coasted on a reputation forged eons before; a leavening of great wines in a sea of dross sustained the reputation but would never be enough to see off this new competition.

Then the weather gods blessed Bordeaux with the splendid 1982 vintage, which fired the starting gun for decades of resurgence, as the wines once again justified their standing, now trading on present glories, with the past as enviable backstory. Today, there has been improvement at all levels, from noblest to humblest, occasioned by the twin forces of better weather conditions during the growing season, and competition from the new world. Put simply, the first resulted in riper grapes coming from the vineyard, while the second prompted better practices in the winery. Mean, palate-curdling flavors, often resulting from a harvest of under-ripe grapes and a slack hand in the winery, are largely a thing of the past.

The clash of cultures between old world and new had hugely beneficial results for consumers in the 1980s. They watched from the sidelines as the drama played out. The crusty guardians of old world values were seen as chauvinists, unable to broaden their judgment parameters, dismissive of anything made outside Europe. Perhaps, but that lazy caricature didn't acknowledge a similar chauvinism among the brash newcomers who sneered at the fuddy-duddy relics, always happy to unleash a barbed put-down. They blew hot and hard; self-effacing they were not. Their wines,

The Damascene Bottle

Every wine lover talks of the bottle that won them to wine, an indescribably subtle flavor sensation that opened a door to a previously unknown world of sensual delight. Mine came while I was still a student.

Like all students I was an indiscriminate drinker, happy to sluice back whatever was on offer. On one occasion we all contributed the cheapest bottle of Port we could find, except for one fellow who brought two bottles of a celebrated vintage. They were like nothing I'd seen before: dumpy and label-less, with heavy blobs of wax sealing the necks. I knew enough to sidle close, offering help with the messy struggle of removing flaky wax and crumbling corks. The ancient liquid was splashed into crude tumblers. It was deep crimson and filled the room with a rich, sweet aroma. The practiced swirl-sniff-sip ritual lay in the future, so a slug was knocked back. The effect was shocking yet delightful, lush sweetness and warmth smacked into the throat, prompting a glorious surge of fumes upward and outward, filling the skull and enveloping the senses. A wan smile followed. I was hooked.

seen in educational terms, had left school early and lived on charm and wits, while Bordeaux had been to university. I loved them both and saw no contradiction in that.

At the other end of the style spectrum came the Warre's 1977, intended to enable a grand millennium celebration to roll on in. It lurched rather than rolled. Y2K was flagged by some as the party to end all parties, while prophets of doom foretold a cataclysm. Both called it wrong. The celebrations barely exceeded those of any New Year's Eve, and the gloomy predictions proved erroneous: no planes fell from the sky, mortgages were not set to zero, and computers didn't crash. It passed with me smoking a cigar, sitting on a rock in Hermanus, South Africa, in midsummer, not a Port-drinking setting.

Thus the stash was saved for other days, and the years since have been punctuated by a bottle now and then, with a few remaining for future delight. Since my first encounter with a dark, grimed bottle Port has never lost its intrigue: the glowing garnet color, the heady aroma, the luscious palate of fruit and spice, treacly rich, warm and welcoming as a log fire on a bitter day. The Warre's is now into its fifth decade and still drinking well, the surge of flavor burnished by time and no longer rampant.

On my first visit to Porto, the city that gives the wine its name, I noted it was as grimy as any old bottle, a tired place, flayed by neglect. If Bordeaux was dowdy, Porto was decrepit, without the same gravitas bestowed by crestfallen grand architecture. That was then. In the years since both cities have been transformed but Porto's metamorphosis is more striking. Even when Bordeaux was in the doldrums an excuse could be conjured for visiting; the same could not be said for Porto. Unless you were in the wine trade it held no attractions. Not any more; now it wears the patina of age rather than neglect. Huge investment has spruced the place up without ruining its character. In the main, greater sophistication rules and a dubious mark of approval for both cities has been the arrival of cruise ships. Those leviathans, with all the grace of a floating apartment block, now disgorge eno-tourists by the thousand, flocking in search of an authentic experience. If they develop a taste for the wines I won't object.

That trio of Rosemount, Chasse-Spleen, and Warre's provided me a springboard for a world of wine exploration that continues with undiminished enthusiasm. I marveled at wine's variety and diversity, its endless landscape, and I still do. I will never know it all, but the desire to see what is around the next corner spurs me on; there is always something new to discover. Realizing you can never know it all frees you from trying; reveling in the journey rather than striving for the destination is what it's all about.

There was a lot to explore. A toe dipped into the mercurial waters of Burgundy resulted in much puzzlement and only hazy enlightenment. The situation has been rectified, though I never expect to solve the puzzle completely. Notwithstanding that, Burgundy hooked me, both wine and place. The wine came first, and I was hugely fortunate to be introduced by a friend to the wines of Henri Jayer, the most celebrated Burgundian winemaker of the twentieth century. That such a bewitching amalgam of scents and flavors could be captured in a liquid was something to be celebrated. My reaction to the best bottles ran to script: silent wonder only broken by swirling, sniffing, and sipping, followed by an outbreak of high-flown descriptors and boundless praise. My first visit to Burgundy some years later resulted in further enchantment. Seeing names such as Gevrey-Chambertin, Vosne-Romanée, and Chassagne-Montrachet at the entrance to those villages brought them to life; they were no longer simply names on labels, with pronunciation traps, such as the silent first "t" in Montrachet. I have walked and cycled the fabled slopes many times since, always marveling at the minute parcellation of the land and the near-religious devotion to the most favored vineyards. Despite crashing disappointments—bottles with warped and twisted flavors that never delivered on their promise—there have been enough soaring delights to stoke the flames of an enduring obsession.

My first taste of proper Sherry was far from enchanting, such was the shock of the new. The nutty savor and chalky dryness ambushed my taste buds; palate recalibration was needed. Some erudite schooling, moderated by a college philosophy professor, was undertaken and I quickly came to appreciate the electric intensity of great Sherry. And that appreciation is continuously bolstered by the criminally low prices producers are able to achieve for even the best examples, which rank among the finest wines on the planet. If you seek value, look no further than Sherry.

Riesling from Germany's Mosel Valley came next. My novice taste buds were thrown again, this time by the counterpoint between fruity sweetness and fleet acidity. Thanks to a wine merchant friend who pillaged his cellar in search of venerable bottles, that deficiency was rectified. Riesling's tentacles radiate out from the Mosel, and I tracked them across the planet, marveling at the grape's ability to retain its identity while echoing the characteristics of every new region. The acidity is like a family crest, sometimes rampant as in Australia's Clare Valley, sometimes burnished, as in Alsace, but always discernible. As with Sherry, though not to the same extent, Riesling remains criminally underappreciated.

Then came the Rhône Valley, moving from south to north, starting with great Châteauneuf-du-Pape from the likes of Domaine du Vieux Télégraphe and Clos des Papes. Again, the palate was intrigued by new flavors; these spoke of warmth and comfort, verging on opulence in the riper vintages. Some magnums from the 1990s remain in the cellar, and there is no hurry to open them. From the north came chiseled, precise Côte Rôtie, as reserved as the lavish Châteauneuf was exuberant. It was hard to grasp that, broadly speaking, they came from the same region, though at opposite ends. The northern wine echoed the tingle found farther north in Burgundy, while the southern one looked to the Mediterranean for added warmth.

I was an early convert to Vouvray, standard-bearer for Loire white wines. The renowned wines of Domaine Huet were my introduction and remain favorites. Penetrating intensity and eternal youth marked them as special, filling a quirky niche for wines that refuse to shout their charms and therefore remain ignored by many. Some antique Vouvrays slumber in the cellar, now fifty-plus years old, rubbing shoulders with their counterpart from Australia, Hunter Valley Semillon, another age-defying beauty.

Next came the twin poles of excellence from Spain and Italy, Rioja and Barolo. At the time the Rioja name was tired, worn out from ceaseless commercial flogging—and many of the wines tasted the same. It enjoyed name recognition but little regard. You had to wade through oceans of dross to get to the good stuff, such as the Marqués de Riscal 1945 and 1964 drunk at the bodega, never having been moved since they were bottled. Seamless elegance and enduring satisfaction marked these as wines that could be rated with the world's best.

Which is not to forget Spain's most celebrated single wine, Vega Sicilia, some bottles of which from the 1960s were carried home from the Barcelona Olympics in 1992. Vega sits in a category of its own, robust and magisterial and only truly expressive after many years in the cellar.

Barolo was a steeper slope to climb, the foothills populated not by featureless wines but by actively aggressive ones that crossed the palate like a runner in spikes. The proper stuff was discovered through one sip of the Monprivato 1985 from legendary producer Giuseppe Mascarello e Figlio. Here was previously unimagined delight, penetrating yet delicate, like the finest aria. Other top names have followed, Aldo Conterno and Vietti being two current favorites.

On and on the discoveries went, including old legends such as Tokaji from Hungary and unheralded treats like Finger Lakes Riesling from upstate New York. More recently I have rejoiced in the revival of old names such as Muscadet and Beaujolais that once looked like they would never escape from the scrapheap of public opinion. Far away in New Zealand, world-beating Pinot Noir is being produced after some wobbly early attempts, though it is now challenged by Syrah as that country's greatest red.

Closer to home Ribeira Sacra and Rías Baixas in northwest Spain give masterclasses in purity of flavor while, closest to home, English sparkling wine is well on its way to world-beating status. Leading that charge is Wiston, perhaps the best of them, which allows Irish wine lovers to enjoy reflected pride from the fact that it is made by an Irishman, Dermot Sugrue.

The Rosemount-Chasse-Spleen-Warre's trio sparked a journey that took me on several laps of the globe. It is a journey that will never end, though recent trips have been conducted virtually. My enthusiasm remains undimmed; there is still much to discover.

2

VINE TO GLASS

Winemaking is tough. Notions of it being a romantic endeavor wither in the face of the winemaker's daily grind. A gifted winemaker striving to get the best out of his or her land in a prestigious region is, in golfing parlance, aiming to win the Masters; this is not a casual weekend round with friends. And the work begins long before a grape is picked.

Conversations with winemakers run to script: type of fermentation vessel, yeast used, fermentation temperature, malolactic or not, time in barrel, type of oak, and so on. These are the building blocks of the conversation, but the one that makes the difference is planting density in the vineyard—the number of vines per hectare or acre. Temperature can be controlled with a flick of a switch, and if pockets are deep the finest oak barrels can be bought with a few taps on a keyboard. But planting density decisions carry great weight, for they cannot be reversed easily and will influence the wine for decades. Once a vineyard is planted, it's virtually immutable. Thereafter the density incurs a significant cost in time and labor—the greater the number of vines, the greater the vineyard work; in a region like Burgundy it is what one generation bequeaths to the next.

There, the benchmark is "meter by meter": one meter between each vine in a row and one meter between rows, giving four thousand per acre (ten thousand per hectare). Anything above that is an indicator of how much tedious toil the winemaker will endure in pursuit of excellence and grants a free pass to the moral high ground of vineyard husbandry. "Great wine is made in the vineyard," is an unchanging truth of quality winemaking, though repetition has ground this tenet to banality, little better than a vacuous political slogan.

Tending to the vines through the winter calls for strength, determination, and skill. The work is relentless and an operation such as pruning is not a mindless chore. Attention to detail is the key: "Cut carefully, don't mutilate," says Thibault Liger-Belair of Nuits-Saint-Georges in Burgundy. Slackness and error in the vineyard will leave an indelible mark on the wine. Quality winemaking begins long before the picture-postcard harvest scene is played out. Not every vineyard worldwide is planted thus, but ten thousand per hectare is considered the baseline for top-notch wine. In Burgundy's Côte d'Or some go higher, to twelve or fourteen thousand, a few go even higher to twenty thousand, a very few go to thirty thousand, in the process condemning themselves to labor that verges on self-slavery. And such diligence is only rewarded in increments, partially when the wine is made and gradually as the years pass, when the wine's class sets it increasingly apart from the products of a less rigorous approach.

The pruning grind lasts for months in all weathers, and there are no magic wands to fast-forward it or other vineyard work, though in the years following World War II supposed shortcuts came in the form of pesticides, insecticides, fungicides, herbicides, and fertilizers that promised an easy route to winemaking nirvana. For a time they delivered on their promise. Bugs and pests and all things nasty were eradicated and, whenever necessary, nature was

Not Chardonnay

It always gets people's attention, telling them that the white Burgundy they are drinking, perhaps a Meursault, is not Chardonnay. Brows furrow, interrogative glances are exchanged. Then I explain that the Meursault gets its identity firstly from the vineyard and secondly from the grape used to make it. It is putting the cart before the horse to describe it otherwise.

Proper Burgundy carries a message from its origin, refracted through the prism of vintage and winemaker. Only then does the grape variety come into play. But wine is increasingly defined by the grape from which it is made rather than where it comes from. "Chardonnay" has been dislocated from the grape it names and now often refers to a wine style. That may work for wines from regions where easy cultivation on massively irrigated land is the norm. But attempting to define Burgundy by grape variety is like trying to measure a person's height using a weighing scale. The grape gives character; the ground bestows identity.

clobbered with more chemicals to keep it in line. Grapes grown with as little trouble as possible was the aim—achieved, but at a cost.

Years of profligate chemical use rendered the soil mute, expressionless like a Botoxed face. The wines echoed the neutered earth: hollow, mere glimpses of what they could be. The silver-tongued salesmen with promise-the-earth patter banked the bonuses generated by easy sale after easy sale. Easy because, to people ground down by a half dozen years of war after decades of pestilence and economic hardship, any promise of relief was welcome. Who wouldn't have succumbed to the sales waffle?

Only after decades of vineyard abuse were heavy-handed treatments reined in and viticulture infused again with good sense, yet a similar malaise soon followed in the winery. In the closing years of the last century advances in winery equipment made possible previously unheralded manipulations. Wine could be pushed and pulled into a desired shape, too often a shape dictated by the market rather than the vineyard. At low price points the new technology allowed wines to be reverse engineered, retreating in precise fiscal steps from supermarket shelf to oil-refinery-sized winery. These fashion-driven manipulations were not as destructive as the excessive chemical treatments—last year's decisions can be reversed this year—whereas the soil takes years to recover. Yet they were indicative of a baleful absence of self-belief, a craven need to temper one's winemaking to the whims of fashion.

That verges on a smug criticism, for it ignores the reality that in a crowded market there is an urgent need to sell and sell again. Primped and polished by the latest equipment, the wines stood a better chance of garnering favorable reviews from the critics, many of whom delivered judgment after barely more than a sniff and sip. Flawless wines became the norm, but they were soulless wines. And the flawless malaise spread right up to some of the most expensive wines, rendering them doll-beautiful, as engaging as a shop window mannequin. Yet some lusted after them, for they can be enjoyed without engagement, the most prized being displayed and stared at in the cellar, shown off for effect.

The vine is cultivated in many beautiful places, usually rugged and challenging. The soil must be poor, for the vine performs best when forced to struggle. It's like physical training—we don't like it but acknowledge its benefits, for the soft life does not produce excellence. In many regions—Portugal's Douro, Austria's Wachau, and France's northern Rhône—the work is exacerbated by the vine's love of steep inclines. It is a hardy yet sensitive plant, able to thrive where few other crops can, delivering sublime wines from slopes that look like moonscapes when bare. Their weathered soil consists of friable stone and little else, yet the vine coaxes riches from it. It can also withstand brutish extremes of temperature, over a span of some 140°F. In a place

14   Wine Talk

like the Douro Valley, the mercury tops 104°F in summer, turning the stony terraces into furnaces. Such temperatures are brain-bending; the heat presses down like a pressure cooker and, while humans can flee to air-conditioned comfort, the vines lie static in its grip.

Winter can bring savage cold. A temperature of 14°F is numbing for humans but still 18° above the -4°F it takes to kill a vine. That's the temperature of a domestic freezer, worth pondering when a rock-hard steak is removed for dinner. Despite such sturdiness, the vine is vulnerable when caught off guard, as in springtime when the delicate buds burst forth. Then, a mere nip of frost can savage the crop in a trice. The winter's labor might be in vain unless extreme measures are employed. In the Côte d'Or, hundreds of small burners are placed at intervals in the vineyards, giving rise to haunting photos of night turned to flickering orange day. What the photos don't show is the precious heat that beats back the chill.

Sometimes nature wrongfoots the Burgundians, such as the night of April 26, 2016, when frost hit with venom, destroying swathes of fragile buds in Montrachet, the most famous white wine vineyard on Earth. In places the damage was so severe there was no hope of making any wine, until seven domaines pooled their surviving grapes to produce two barrels, where two dozen is more normal. In time, this will become one of the rarest, most expensive and sought-after wines in the world.

The rising sun compounded the damage in 2016, its rays magnified by the globules of ice on the buds so as to burn them. The following year every bale of hay that could be found was set alight at sunrise to form a protective cloud. It worked—and caused havoc on the roads, so heavy was the smokescreen. For a while, hay bales stacked beside vineyards in springtime became a common sight but the practice is now forbidden.

Once through the spring frosts there's little relief before summer hailstorms threaten. It is difficult to describe their vicious intensity, the shock factor exacerbated by their rapid arrival and departure. They pummel and are gone, the sky-borne blitz wreaking destruction in minutes. Various measures such as protective netting are used to combat the threat, but the cost and questionable efficacy limit their employment. Curiously, the best defense remains passive and comes by way of extraordinary fragmentation in vineyard holdings. Most Côte d'Or domaines comprise a host of scattered plots, so the localized nature of the hailstorms means that severe damage in one is not repeated in all others. It's scant consolation, but with survival balanced on a knife edge that can make a critical difference. Otherwise, the gods can be implored and fists shaken at the sky, measures that remain as ineffective today as before. Winemakers are largely helpless in the face of nature's ire.

Depending on where it is cultivated, the vine presents many faces to the world, from shipshape to ragged. In patrician Bordeaux it is trained into endless rows that cross-stitch the land in mesmerizing patterns. The prices charged for the wines justify the cost of such attention; if they were people they would be neatly coiffed and clean shaven. At the other end of the spectrum come the bush vines, tousle-haired hippies of the vineyard that grow untrammeled by the wires and posts that guide their Bordeaux cousins to excellence. Grenache is often cultivated this way and the wines are flamboyantly flavored compared to Cabernet Sauvignon's well-ordered flavors.

Regardless of location, the vine is rigorously tended; the level of manicure varies, but nowhere is it allowed free rein, not even the shaggy bush vine, because it would sprawl like a badly planned city, its grapes as featureless as identical suburban dwellings. It is only by rigorous cultivation, yoked to a winemaker's philosophy, that the vine yields its best. Otherwise its potential lies dormant, never released by harnessing. Planting it on easy, fertile soil results in a big crop of slack-flavored wines. Make it suffer and you get quality.

The vineyard workers also suffer, enduring relentless labor as they tend their charges, with each region carrying its own challenges. The Douro's baking-in-summer-freezing-in-winter terraces are akin to a treadmill, to be climbed and descended, climbed and descended. To boot, the traditional slate stakes used there demand a weightlifter's strength for handling. They are rough cut and dark, hot to touch in summer with graphite gray facets glowing in the sunlight. I once carried a chunk home in my suitcase, the longest piece I could fit diagonally, wrapped in dirty shirts to minimize the clatter. Hefting it about was difficult enough; it's hard to imagine a truckload.

The challenge in Sherry country is different; lungs and legs are untroubled but the eyes screw up from the dazzle of the region's *albariza* soil—pale gray, dusty, and glaring in sunlight. The breeze sets it swirling—leaving the vines ghostly, as if dusted with talcum powder. After that the dark of the bodega brings relief, the eyes unravel and adjust to the pleasant gloom, while the nostrils twitch on the savory aroma of wood, earth, and wine.

In addition to slope and soil, winemakers are challenged by wind, and when it blows without respite measures must be taken to mitigate its pummelling. On the island of Pantelleria, a tiny blob of Italy that lies closer to Tunisia than Sicily, the vines hunker in individual "craters" dug for them. Each sits snug in its hollow, growing wide rather than high to duck the wind. Further protection comes from the dry stone walls that add an ancient look to the landscape. The system of vine training is known as *vite ad alberello* and has been accorded the status of "intangible cultural heritage" by UNESCO. Such acknowledgment embellishes its historical significance but does nothing to ease its challenge. Winemaking on Pantelleria comes with a laborious price.

The contrast with the vineyards of Galicia in northwest Spain could not be greater. There, the vines are trained up granite posts onto overhead pergolas, sharing the ground below with other crops. Galicia is a verdant land far removed from the Spain of stereotype: the baking central plateau with sun-bleached sky, the robust red wines and hearty meat dishes, the windmills of La Mancha where Don Quixote roamed. In Galicia seafood takes precedence over meat and Albariño is the wine, made from grapes grown in garden-sized vineyards that make the blocky granite posts appear incongruously large, looking like they were crafted to support a railway line.

Great Grape: Albariño

Well regarded by the cognoscenti and now the darling of the chattering classes too, Albariño is Spain's riposte to the ubiquity of Sauvignon Blanc and Pinot Grigio. And that latter association is not to suggest it is a chinless wonder, far from it. There is character here, allied to an easy-to-like appeal that hooks newcomers from their first sip and, crucially, keeps them engaged right through to last swallow.

Albariño's homeland is Galicia, known by many as "green" Spain. The wines are as fresh and immediate as the breeze off the neighboring Atlantic Ocean; they call to be quaffed without delay. Yet they age far better than many realize, shedding exuberance while gaining creamy nuttiness, satisfying gravitas replacing youthful raciness. Galicia might also be called granite Spain, for the stone is seen everywhere. As with Muscadet well to the north, there is a hand-in-glove fit between Albariño's vitality and the sea's bounty, harvested from the chaotically fractured coastline. Eons of pummelling by the Atlantic have rendered it so—in sharp contrast to the wines' smooth progress across the palate.

In these challenging regions the quirks of geology and geography endow potential that must be mined like a seam of gold. The wines, compared to those from industrial-scale vineyards, are better, more characterful, and, ultimately, more rewarding. Just as the vineyards contain potential so do the wines, frequently locked deep inside. A sum of parts goes into the bottle, there to be burnished by the years, eventually developing into something greater, something barely glimpsed at bottling.

Winemaking that starts at the point of sale rather than place of production can be done anywhere; the wine is of nowhere. Cost is the decisive factor, and industrially produced wines seek out the places where grapes can be grown with clockwork regularity. Rigorous quality control then leads to dull homogeneity, facsimile wines of feeble identity embellished with cod tales from the marketing people. Eye-catching labels distinguish one from the other, but if the labels were swapped, would anybody notice? It is not unknown for some everyday wines to have two different labels, one for direct sale to consumers, the other for restaurants, to stymie attempts at calculating markups.

Many mass-produced wines need some stiffening to bolster their character, and for reds an additive called Mega Purple does just that. It sounds like Darth Vader's cousin yet it is harmless, acquitting itself quietly as the additive that dare not speak its name. It's a concentrate whose primary purpose is to deepen color and is made from Rubired, a *teinturier* grape, meaning its skin and juice are both red. Aside from darkening wan red wines it bolsters them in other ways: a scratchy texture is smoothed toward plumpness, a vacant flavor is coaxed toward richness. It is like "staging" a house in advance of a sale; at a stroke the wine appears to jump in quality.

Asking a winemaker if he uses Mega Purple is akin to asking a lady if she uses hair color. And it is more than color; it's magic potion for cheap wine, giving the appearance of substance, but it is piling superstructure on frail foundations. The wines resemble a film set—all facade with nothing behind. Winemakers can hardly be blamed for deepening color, for consumers frequently choose on appearance—think straight carrots and blemish-free tomatoes—and for some reason depth of color is seen as an indicator of quality. Buying on price alone exacerbates the problem.

The darkness of some "red" wines today suggests "purple-black" as a better descriptor. Some could probably be used as ink. That stygian color should not be confused with the deep violet of a great young Bordeaux, for instance; it is dark but also vivid. Nor should the shimmering crimson of Burgundy be forgotten—its pale glow is still prized by those who know better.

Adding Mega Purple perfectly legally is not much different from the adulteration of old, when rugged Algerian wine was used to bolster weedy French ones. Curiously, as the Mega Purple is heaved into red wines, there's little to be gained by bleaching white wine to watery clarity. Pale and wan doesn't cut it; deep and bold is what's needed. Despite such treatment, industrially produced wines reach a consistent standard, top of the class in quality control tests. They have replaced their scrawny predecessors that scarified the palate and instead they lull it to somnolence with monotone flavors. Flawlessness is their fault. At the end of the last century they conquered the world, brash and branded to the hilt, flooding into Europe from Australia, Chile, California, and elsewhere.

The workhorse wine regions churn out a vast bulk of wine. They are not photogenic and are chosen solely for ease of production-line grape-growing; reliability and consistency are their virtues. Australia and California are home to some of the most productive. In the former the vast inland region of Riverina, centered on Griffith, presents a vineyard scene of dreary regularity, the eyes droop at the lack of stimulus, as do the taste buds when the wines are sampled. This is monotony writ large—maximum yield with minimum effort.

The picture is similar in California's Central Valley—there's none of the cosseting, the treating like favored family pets that goes on in the likes of Montrachet, or at Petrus. The valley is no place for spoiled-brat vines. It is hot and dry and churns out some three-quarters of California's grapes and, as with Riverina, the "challenge" is the lack of rainfall. Irrigation is an absolute necessity, so water is pumped in and ladled onto the vines and they yield and yield and yield ... until the water runs out. It hasn't done so yet but these unnatural environments for grape-growing place a huge strain on supplies. If the water runs out, the flat and arid vineyards will disappear. Come vintage time, monster machine harvesters trundle through the vineyards, looking like mobile oil rigs. The vines shiver as they pass, the ground rumbles, and the air pulses; standing close is not for the fainthearted.

Huge crops ensure the flavor is stretched thin, like diluting paint to make it go further. The wines are a blank canvas upon which identity is imposed in the winery. It is like a chef working with battery-farmed chicken—all sorts of herbs, seasoning, and spices have to be added to give flavor. The winemaker utilizes equivalent adjustments to churn out squeaky-clean wines aided by all the winery gadgetry, but precision has triumphed passion, adding a gloss to obscure the vacant spots.

Once the new world had colonized the bargain basement, ambitious producers turned their gaze to the high ground—populated by Bordeaux's classed growths and Burgundy's elite *crus*. Success in that rarefied world was harder won, entrenched opinions were slow to yield, but help from an unlikely quarter was at hand and it came in 1976, the United States' bicentenary year.

The wine tasting, since named "The Judgment of Paris," was organized by an English wine merchant based in Paris—not, at first glance, the sort of person likely to rock the boat. Steven Spurrier assembled some of the finest French wines, and Californian equivalents, and arranged for them to be tasted blind by an esteemed group of French tasters. All was set to rubber-stamp France's superiority, except the Californians hadn't read that script and ended up being preferred by the "home team" tasters. Spurrier delivered a boot in the derrière to the Gallic wine establishment and, by extension, all of the patrician old world whose smug conceit was stripped away by this dose of reality. It was painful and badly needed.

Similar tastings have been held since, but ceaseless flogging has debased the coinage and there is now a hollow ring as the old world is upstaged again by overseas cousins. With each iteration they become less interesting, the rabbit can't be pulled from the hat ad infinitum. These events have served their purpose and, in truth, they are a crude yardstick by which to assess wines, akin to standing people in a lineup and saying tallest is best. A "French Paradox" also applies to such tastings. No matter how often France is trounced, it doesn't damage the reputation or affect the price that

Will It Age?

This question—asked in dubious tones—was much to the fore when the high-end, ambitiously priced wines from the likes of Chile came on the market. I asked it myself, specifically of Almaviva, the luxury Bordeaux blend made by the joint venture between Mouton Rothschild and Conch y Toro. Tasting the 1999 vintage—roughly an 80:20 blend of Cabernet Sauvignon and Merlot—from barrel on a visit to the winery that year, I was impressed by its scale and structure but my notes concluded, "will it age?"

A resounding "yes" was the answer on my return to the winery nineteen years later. The building blocks of tannin, acid, fruit, and alcohol had melded to smoothness, emulsified by the years. Where once they stood separate, like the paints an artist uses, now they were almost indistinguishable, a delicious amalgam of richness and restraint. Time had added harmony to the original melody; its passage is the true test and cannot be bought or accelerated. Patience is needed.

the top wines command. The grandees of Bordeaux and Burgundy remain the most sought-after and expensive. They have been joined at or near the top of the pile by the likes of Almaviva and Seña from Chile, Penfolds Grange from Australia, and Opus One and a slew of others from California, but they haven't been displaced. The world has shifted slightly on its axis, but only slightly. In addition, these events are no longer needed, for such is the quality of the best new world wines today that they deserve better than a beauty pageant to display their charms.

To repeat, quality winemaking is tough work, not for the weak-willed. A region like Italy's Piedmont, where the vineyards wriggle and twist on the jumbled hillsides and everything is small scale, is peopled by farmers whose crop yields wonderful Barolos and Barbarescos. Think of them as farmers rather than winemakers; it's a more accurate description of their daily grind, there can be no slacking, no lazing on the terrace, for the vine is not self-catering. After months of effort harvest approaches, vineyard work ceases, and the waiting for optimal ripeness begins. The weather forecast is scrutinized like never before—dare they wait for another increment of ripeness or harvest before imminent rain?

Once harvest starts, life becomes a frenzied blur of activity, with every ounce of effort expended in getting the grapes safely to the

Great Grape: Nebbiolo

You're never sure which Nebbiolo will show up until you open the bottle: singing nightingale or croaking frog. It's the bipolar grape that enchants and frustrates in even greater measure than Pinot Noir. When on form Nebbiolo delivers the most beautiful vinous aria, and when not, it's a severe libation, razor wire made wine. Granite tannins and screeching acidity see to that; what is needed to counter these twin terrors is a combination of sufficient fruit and the passage of time. Only then will Nebbiolo blossom into storied magnificence.

If this isn't enough to keep it forever from the mainstream, its aversion to travel will. When planted beyond Piedmont, Nebbiolo pines for home by losing its signature scent and allure; in a foreign land it sulks, scaring away otherwise adventurous winemakers. And on home ground, if made cheaply and without diligence, it is sinewy, with a patchy flavor. There's hardly another grape variety with such a chasm between best and worst, and that makes it the insiders' choice. It is the most overlooked great grape in the world, adored by aficionados and disliked by others. A great Nebbiolo is haunting and seductive. Almost none can match it. Seek it out.

winery. When completed, all the work is concentrated there, the vineyards fall silent, and the foliage brittles in the chill evenings. The uniform green leaches away, replaced by a dazzle of copper and russet, ochre and gold, carpeting the land like a Persian rug. The winemakers are busy, but the preharvest anxiety eases; everything is in their hands now. The quality of the grapes, cast in stone by the weather and their own husbandry during the growing season, dictates the potential of the wine; the winemakers' task is to harness that. It is a limit beyond which they cannot go, but must not fall short. The most talented always test the limit, wresting flavors from the grapes that less gifted peers can only marvel at, particularly in weak vintages. Meanwhile, autumn fog has descended outdoors and the forests stir to the sound of truffle hunters and their dogs, seeking out the region's "white gold."

The picture is the same in Burgundy, though the hunters there are in pursuit of wild boar, and once the fermentation is complete, with the wines settling in for a few months in barrel or tank, the vineyard cycle begins again. There are few visitors to witness the silent grind. The cycle tourists who swarmed about during harvest, capturing everything on helmet-mounted cameras, are gone. The odd wine professional, darting hither and thither from appointment to appointment, barely notices the slowly moving figures in the vineyards as he struggles with a grim schedule, conjured back home at the desk. Timings turn notional, frustration and fatigue set in, but then the day is rescued by a generous winemaker keen to share a venerable bottle after a long vineyard shift.

Winemaking itself is simple; it is not wizardry presided over by Bacchus. Fermentation is a well-understood process whereby the action of yeast turns sugar in the grapes into alcohol, with carbon dioxide as a by-product. Nevertheless, it is worth pondering the transformation, for it reorders the flavors, disassembles the juice, and reassembles it as wine. The flavor transitions from juicy and sweet to juicy and savory, the sweetness replaced by alcohol. The fruit is still there, but its flavor register has been altered. It is hard to think of non-sweet fruit, for it is sweet almost by definition, yet even with its sweetness removed it is not hollow thanks to the alcohol. Much remains the same but the essence is different and that transformation never loses its fascination.

For a year or two, depending on style, the wine is nursed along, every barrel frequently tasted to monitor its development, and topped up to avoid spoilage. This is the *élevage*, the "raising" of the wine. It's akin to raising children, which is why winemakers never say which of their wines is their favorite. When bottling time comes, many small- to medium-sized producers use truck-mounted, mobile bottling units rather than invest in a piece of equipment only used for a couple of weeks each year, being shunted to a corner of the winery the rest of the time. At the industrial end of the business the bottling line is a permanent installation, in use year-round and vast in scale. The largest I've seen was in La Mancha, Spain, and comprised five separate lines housed in a "football field" warehouse, all operating at mesmeric speed and efficiency. Not all wine is bottled at the winery; much is shipped in large bladder tanks and is bottled closer to its point of consumption than production, a detail often missed by consumers.

The type of closure used to seal the bottle is one of the hottest debates in wine right now. There was a time when the cork industry

Blue or Black?

Some wags spin a yarn that in the less-than-sophisticated Dublin dining scene of the 1970s—with the eclectic landscape of today lying well in the future—in some restaurants, the wine offering was blue or black, not red or white. This referred to Blue Nun or Black Tower, twin pillars of vinous sophistication in too many parts of the world back then. Both were Liebfraumilch, an anodyne brew of dubious merit. It was gently confected and could be sluiced back without effort, just as Pinot Grigio is today. Elevator Muzak never came blander.

had a virtual monopoly, churning out billions of corks of varying quality. Consumers' too-ready acceptance of wines ruined by poor corks gave the producers a free pass and only when their customers, such as winemakers in Australia and New Zealand, rebelled and switched to screw caps did things begin to change. Today, thanks to the white heat of competition, cork quality is much improved. It might even be argued that screw caps saved the cork industry from lazy decline, though I doubt any letters of thanks have been penned to that effect.

However it gets into the bottle, the wine is now a bird flown from the nest, and it is the job of the sales and marketing people to get it into your glass ahead of all competitors. They'll have a hand in label designs—from sober to zany—all crafted to send an irresistible buy-me-drink-me message. Packaging is far more important than consumers care to admit. Producers go to great lengths to get their wine noticed ahead of the scores of others it will be rubbing shoulders with on the retail shelf. The bottle shape and size are important, but paramount is the label. Some are essays in simplicity, such as those of Benjamin Leroux in Burgundy. Others are a riot of color, none more so than the "Porta 6" label of Vidigal Wines. Each draws the eye by different methods but with the same purpose. We drink with our eyes—try decanting a favorite bottle of wine into a plastic water bottle and see how its attraction wanes. Retailers play their part too, with discounts and quantity buys, including risible "half-price" offers that test the bounds of credulity. We wade through the maze, always in search of the next good glass.

3

BUBBLES

A bloodred, foot-shaped piece of rubber protruding from the soil caught my attention so I tugged it free. It was the insole of a sneaker and I stood there with a fly-in-my-soup face, baffled. Then I noticed speckles of blue and black and green. Scuffing with my foot revealed shards of refuse sacks, a Bic pen, items of household garbage. Nothing prepared me for these finds on my first visit to the Champagne vineyards, least of all the publicity material that the promotional body, the *Comité Interprofessionnel du vin de Champagne*, turns out. Champagne's grubby secret is that the vineyards were used for years as a dumping ground for Paris's rubbish, and it's a secret so at odds with the expensively crafted image that it is never spoken of. In truth, it is fading from view, for the dumping has ceased, and the vineyards are returning to normality. The reason it happened at all was that historically, it was mainly food waste and acted as fertilizer rather than rubbish.

* * *

27

Champagne. Perhaps the most evocative word in the world of wine, and beyond that insular world too. Everybody knows what a "Champagne lifestyle" is; the term conjures images of celebration and luxury, pomp and grandeur. The French, historically noted for foot-in-mouth wine marketing, have always got ten it right with Champagne. It's slick and sophisticated, using glamour and association with success to burnish the image. And having achieved this, Champagne defends itself belligerently against any who seek to piggyback on its success, sending in the legal rottweilers to bully small prey should they usurp the name "Champagne" in any way. *Les Champenois* patrol the wine world to ensure nobody makes "Champagne" anywhere but in the region itself. They have managed to stop the use of the term *méthode champenoise* on the labels of other sparkling wines. In one famous incident, they stopped the village of Champagne in Switzerland from using its own name on its wine. Their power also reaches beyond the wine world and any product such as a perfume that dares to call itself "Champagne" will feel their wrath. They were rocked back on their heels, however, in July 2021 when it was decreed in Russia that all imports should be labeled "sparkling wine." There was predictable outrage, much huffing and puffing, with the whole kerfuffle reported on in breathless prose under clickbait headlines. The story flashed around the world, then almost immediately ran out of legs, when it was realized that only the back label would be thus sullied, and the precious front label would still announce clearly: "Champagne."

This desire to protect their cherished wine from the sincerest form of flattery is understandable. Yet the ludicrous lengths they go to suggest a malaise founded on insecurity and fueled by a craven need to swat away all perceived challengers, no matter how feeble. Such schoolyard bully behavior wins them no friends and yet, curiously, it doesn't lose them customers, so we can expect it to continue, further bolstering their unchallenged position and its attendant smugness.

Extra Quality, Very Dry

My introduction to Champagne was a full-immersion baptism at Henley Royal Regatta when still a student. Knocked out in the first round of racing, I was long on time and short on cash, so was stretching a tepid pint of beer across a sunny afternoon when I fell among a coterie of Champagne-drinking revelers keen to share their largesse.

The Champagne was Bollinger. "Extra Quality, Very Dry," announced the label, doing little to prepare my palate for the flavor shunt from meager beer to this, a drink of verve and penetration, austere not astringent, exuberant yet reserved. The flavor unfolded like a wave coming into shore and then lingered before ebbing gradually. Another sip, another wave. I was hooked and my new best friends were determined to keep it that way. They were in buoyant mood, but there was no Formula One spray-about nonsense. The afternoon passed in a trice.

All of which is a shame and a distraction from everything that is wonderful about Champagne, for this is not just the wine of shallow celebration. There is pure delight also, a joie de vivre lift of the spirits that matches the rush of bubbles as the cork pops and the first glass is poured. And that delight prompts serious appreciation of the best examples: initially the exuberant flavor, then the deeper layers of complexity, then the sumptuous aftertaste only perceived when the initial pyrotechnics subside. These wines compare with the world's greatest and deserve to be treated as such. Champagne is a many-faceted wine, and those charged with its promotion failed to champion that until recently; theirs was a one-dimensional mantra centered on celebration sans contemplation.

A good-natured gathering is the most common scenario for Champagne, though raucous conviviality is not ideal for anything other than a basic wine. The better ones warrant thoughtful presentation, starting with the glass. Champagne glass shapes widen and narrow almost as hemlines go up and down; the horizontal, saucer-like coupe gave way to the vertical, tall flute, the first anathema to appreciating the rise of the bubbles, the second favoring the bubbles to the near exclusion of the aroma. Both still have their advocates, but the shape du jour has gained amplitude in every dimension and is little different from a conventional wineglass, which is a bit pedestrian for me. The coupe and the flute have an element of frivolity, whereas this is almost anti-frivolous— it should at least descend to a point at the base of the bowl to prompt a steady surge of bubbles. Twist the stem to set them spiraling. The best glasses should allow for appreciation without stifling enjoyment.

Whatever about glassware; the custom of serving Champagne only as an aperitif must be broken. It verges on a taboo and sees Champagne considered as a one-trick pony, corralled into the half hour before the meal. Or, catastrophically, it is reserved for meal's

Take Five

More than any other wine, Champagne appeals to all five senses, beginning with the sound of the cork. Champagne reps used to say it should sound like a maiden sighing, a racy allusion that would get them canceled now. Then the eye is caught by the lazy rise of the bubbles and after that an enticing fragrance greets inquisitive nostrils, often baked apples. The apples continue onto the palate in a host of iterations, under-ripe in cheap, malnourished tipples; sumptuous and abundant in better examples. Finally, Champagne's touch on the tongue is more memorable than any other wine and is an integral part of the enjoyment. In the best wines it's a soft caress, in lesser ones an aggressive stab.

end, frequently alongside wedding cake as the newlyweds are toasted to the heavens. This is a gustatory crime, as crisp wine and luscious cake collide on the palate. They arm wrestle for supremacy until the weight of sweetness carries the day and the Champagne cedes victory. Even without cake, reverting to Champagne after a meal is a challenge, for it is a livener, an uplifting wine that stimulates the senses and sharpens the appetite. In crude terms it is the energy drink of the wine world and as such is at odds with the hour of repletion after a meal. Unless it is lunch and dinner beckons.

Is Champagne wine? Yes, a thousand times yes. In a sense it was wine before it was Champagne, and there lies an awkward piece of history: sparkling Champagne, as opposed to the still wine for which the region was known in the seventeenth century, first saw light of day in England, not France. English glass, fired in coal-burning furnaces that were hotter than French wood-burning ones, was stronger and hence could withstand the pressure from within. Shipped in barrel, the wine was then bottled and whether it ended up sparkling or not was a haphazard occurrence. Initially, this was most likely regarded as a fault and not the prized quality of today, but bubbles eventually won and are now examined and commented on—for persistence and size and quality of the frothing mousse. A steady stream of small bubbles is best.

Today the process of creating those bubbles is carefully controlled, and a little sugar and yeast is added to the wine as it is bottled to prompt a secondary fermentation; the resultant carbon dioxide is trapped and dissolves into the wine until it is opened. Before then, however, the sediment of spent yeast cells needs to be removed. With the bottle upended, the neck is frozen by immersion in a bath of sub-zero brine and then the plug of sediment is expelled by popping off the crown cap used for this part of the process. A deft hand is needed to do it with minimal wastage and when done at speed it sounds like rapidfire pistol shots. Only then is the bottle sealed with the familiar mushroom-shaped cork, though that starts life as a plump cylinder, being reshaped by constriction in the bottle's neck.

That is the Champagne process or *méthode champenoise* taken at a gallop. It is labor-intensive and time-consuming and some stages require great skill and experience. The most critical is the blending, the first step in transforming unremarkable raw ingredients into the jewel that froths in the glass and dances on the palate. Where the alchemists of old failed to turn base metal into gold, today's Champagne blenders succeed with equally dull base wines, elevating them to excellence. The measure of their success can only be appreciated by tasting those base wines.

I once took part in a blending exercise and the memory remains sharp; it is not for prim palates. In a functional laboratory dozens

Size Matters

Champagne bottles come in a bewildering range of sizes, though the standard 75 cl bottle and the 150 cl magnum are the ones regularly seen. Going bigger, bewilderment sets in, thanks to a surfeit of biblical names, starting with the Jeroboam, named after a king of Israel and equivalent to four bottles. It is the largest bottle produced in commercial quantity.

Then follows a quintet of Old Testament notables: Rehoboam (6 bottles), Methuselah (8), Salmanazar (12), Balthazar (16), finishing with the mighty Nebuchadnezzar named after the greatest king of ancient Babylon and holding twenty bottles. There are even larger bottles, a rogues' gallery of grossly proportioned howitzers. There is no grace to their curves; they might make a splash for publicity purposes or they might explode. Large-format bottles produced in small numbers are not as stable as the standard sizes, and are sheathed in clear plastic in Champagne cellars in case of accidents. Stand well back.

of sample bottles identified by alphanumeric codes awaited the blenders' attention. I set out boldly, only to have my taste buds scarified by rivers of acid. These wines screeched, angular and raw, and left my mouth strangely parched though the saliva flowed in torrents. They tasted nearly identical, differentiated only by the intensity of the stiletto acidity, as hard on the palate as a car alarm on the ear. I mixed a few together and presented my blend for the boss to dissect forensically. There were weak spots, lack of breadth, shortness of length. Each deficiency was made good by the addition of a carefully selected sample, until the remedied blend was returned to me. It was remarkably transformed, as a chef combines ingredients to yield a flavor undreamt of initially.

A wine into which so much effort is put, a wine so celebrated, deserves respect, but paradoxically it is the most abused: Formula One victors spray it about; legendary beauties bathe in it; and ardent lovers drink it from slippers, all of which are partially excusable. What is unforgivable is when the producers demolish their own product, having said how much trouble goes into making it. The Moët et Chandon Champagne fountain, where the wine is poured into the top glass of a pyramid until it overflows into the lower glasses, is a Gatsby-esque travesty that turns the wine into a fairground attraction. If people don't regard Moët et Chandon highly, Moët shouldn't get annoyed, they don't do so themselves. Are they not proud of it? Is it made in such vast quantities that it doesn't matter?

Until recently, most of the effort was expended in the cellar, which was where Champagne was given its character. The vineyards were little talked of; their input was submerged under the brand identity. The wines tasted of who made them, not where they came from. The blender's art is paramount in Champagne where, despite the recent trend toward identifying a wine's vineyard origins, the brand remains king. Consistency is all, the house style a bottled mission statement: reserved and vigorous Bollinger, full and foursquare Krug, balanced and harmonious Pol Roger, elegant and correct Ruinart, soft and fruity Laurent-Perrier, fine and full Roederer.

Does Not Rhyme with Jug

The house of Krug remains the alpha male of the Champagne world, but in recent years Krug aficionados were heard to mutter into their bubbles that the uncompromising style had softened a little, weight sacrificed in favor of easy attraction. Where, they pined, was the magisterial style of old, at once forbidding to neophytes and compelling for aficionados? Even the label turned lacy and lissome, a frivolous essay at variance with the traditional house style. The advertising expended a lot of effort in telling—a little too earnestly—how good the wine was rather than letting it speak for itself, as previously.

More recently I have wondered if the style of the flagship Grande Cuvée has changed again, flipping back and firming up as if in search of past glories, prompting me to note: "An iron fist in need of a velvet glove. Big, rich stentorian stuff, very much in the savory register. Great depth and length, not much dazzle. A bit shouty." The label has reverted to a more somber register, giving hope that the style will settle once more after these oscillations. Throughout, there has been one constant: it's always pronounced "Kroog."

The brand has been the gift that keeps giving for so long that its creators are unlikely to undermine it. Today's Champagne innovators are the smaller players, the mavericks with little to lose from disrupting the status quo and who, with a glance toward Burgundy, see the character that individual vineyards can bring to the wines. Change is in the air and the bigger names are jumping on the terroir bandwagon, mirroring the beer industry's response to craft beer's disruptive influence. Champagne's cast-iron consistency achieved year after year is impressive, but consistency's bedfellow is boredom. Champagne should deliver wine's endless variety and diversity. Too much has been done to stamp that out.

Effort is being extended back up the production line and out into the fields, where the land is now a cherished asset, not a flogged workhorse. The practice of rubbish dumping was discontinued in 1997. Champagne is coming late to the terroir party, where the Burgundians have been for centuries, dividing and subdividing their treasured hillsides into arcane tangles. Even the Australians, wedded as they are to varietal labeling, are moving toward more site-specific identification of their wines. *Les Champenois* might argue that they too have been there for eons, well aware of the attributes of Chardonnay from this slope or Pinot Noir from that hillside. Yes, but it was information they kept to themselves, hidden behind the brand's facade.

Blending doesn't only occur across multiple vineyards and grapes but also across vintages, another factor that dislocates Champagne from its origins, making the cellar its place of birth. Blending irons out inconsistencies from year to year but, just as the trough of a bad year is avoided, so the peak of a great one from a favored vineyard is missed in all but a clutch of Champagnes. Apart from celebrated exceptions, such as Philipponnat's Clos des Goisses, vintage-dated, site-specific Champagnes are rare, though we are likely to see more of them in coming years. More attention is now being paid to vintage Champagne, allowing the weather a wild-card role to roast or batter the grapes. Every vagary leaves its mark. Climate change, however, is easing the challenge of growing grapes in a marginal region, prompting further shift in the balance between vineyard influence and cellar control.

Change is in the air. The old divide between the thousands of farmers who grew the grapes and the producers who bought them is blurring. Numerous growers are now making their own wine, spawning a rivalry among merchants to source a never-before-heard-of Champagne with which to impress their customers. The style of Champagne is also changing, courtesy of hotter summers delivering riper grapes lower in acidity. That's a sweeping generalization, but it holds water. In the past the *dosage*, the

The Third Man

The principal grapes of Champagne are Pinot Noir, Chardonnay, and Pinot Meunier—the first for body, the second for grace, and the third because there is a lot of it. With some exceptions, all Champagne is made from this trio in some combination. The first two are the aristocrats while Pinot Meunier, of which far more is planted than is talked about, is the workhorse. However, recent developments suggest that Meunier's day has arrived. There was a time when it was barely mentioned, though Krug always noted its significant role in their blends. Now, with a splintering of Champagne's monolithic face, all producers are seeking points of difference. Wine with Meunier as the star, if not the sole player, is one way of doing so. Prepare to see and hear more of it in the future.

sugar-wine elixir added to every bottle, was akin to vinous makeup, smoothing the acidic wrinkles, but less is needed now. The new moral-high-ground Champagne is *dosage zero*, but without the mild slick of sugar these laid-bare wines can be bleak and scrawny.

Apart from some blips, worldwide demand for the fizzy stuff galloped ahead over the last century. Today, annual shipments usually exceed 300 million bottles, with some recent slackening because of COVID. That figure is put into context by the 100 million shipped half a century ago. Booming sales filled bank accounts in the cities of Reims and Épernay until an unforeseen problem was encountered—the supply of grapes could not keep up with demand. Historically it was always possible to plant more vineyards, but this door was now closed; every scrap of land in the strictly delimited region was planted. Producers dug into their reserves to meet spiraling demand. Large reserves are an important piece in the Champagne jigsaw, but they struggled to cope with the new millennium's thirst.

In the early years of this century it was proposed that the region should be expanded so as to cash in on burgeoning demand. Every manner of survey, including satellite imagery, was employed to identify the correct sites; a lot was riding on this. Any scrap of land sprinkled with fairy dust and raised from serfdom to aristocracy would soar in value. One wonders at the reaction if those satellites had strayed and discovered the south of England was perfect for Champagne production.

This gave the cynics a field day. They watched in glee as the Champagne authorities' suave touch faltered, when spinning their proposal to a raised-eyebrow world. At a stroke, it made a mockery of their argument, pedaled ad nauseum, that for a sparkling wine to be Champagne it had to come from the delimited region. Only geographical provenance could guarantee authenticity, but now geography was to be rewritten. The proposed expansion would be an ugly-duckling-to-swan exercise of breathtaking arrogance, with greed as its handmaiden. Shortly afterwards the world economy stalled, softening the expansionists' cough, and the heated debate cooled in the face of more pressing concerns.

It is time for the authenticity argument to be toned down, with more energy devoted to improving quality at the lower levels. Too much mediocre Champagne is foisted on the world every year, wine that is kept on life support by having the magic "C" word on the label. Authenticity is one thing, but quality is another. The latter should be given precedence. At its best Champagne is one of the finest expressions of the winemaker's art, a wine of boundless energy supported by more nuance and complexity than many wine lovers realize. It can age superbly too and fully deserves its place in the pantheon of the greats—shorn of arrogance, that is.

Despite all the hokum pedaled by the big brands about their bubbly, I remain a sucker for Champagne, and the region is only forty-five minutes by high-speed train from Paris, which makes a day trip for lunch possible. But stay awhile, there's much to see, including the cathedral of

Notre-Dame de Reims, where French kings were crowned for hundreds of years. In a country replete with ecclesiastical monuments, this soaring edifice still sets jaws dropping. It sits at the heart of Reims, the region's capital, and was almost destroyed by shelling in World War I. No visit to the city is complete without an exploration of its interior; the stained glass windows by Marc Chagall are a highlight. Eschew the babble of the guided tour in favor of a silent hour to savor the splendor. You can pick up the facts and figures later from a guidebook.

Reims presents an insurmountable pronunciation hurdle for English speakers. "Reems" is out-of-the-park wrong. The solution is simple: remove the "F" from "France" and you have it—"Rance." Start the "R" in the throat, give it an extended roll and you'll pass for a local. When in Reims don't go in search of a Champagne cocktail. We live in a cocktail-crazed age, populated by ever more

Ponder This

It can be difficult to grasp the fact that Chardonnay, progenitor of the great white Burgundies, and Pinot Noir, which serves the same role in red, combine in Champagne to yield the greatest sparkling wines of all. The contrast between the still wines made from this pair, and their sparkling cousin produced well to the north of Burgundy's Côte d'Or, is little short of startling. The Burgundies stand as the finest white-and-red pair from a single wine region anywhere on Earth, a fact that doesn't suggest them as partners in an equally compelling sparkling wine.

How strange it is then that the Champagne region, despite utilizing the grapes of Burgundy, is only now catching on to the Burgundian model of viticulture, where the soil is treasured rather than trounced by vignerons. In fairness, Burgundy has only recently rediscovered an appreciation of its greatest asset, and it is encouraging to see that Champagne has started out on that road. May the journey be quick.

lurid creations, but that's no excuse for ruining this great wine. The sorry combination favored over the decades combines the wine with sugar, brandy, and angostura bitters, a ragbag of flavors. Each ingredient concusses the wine, with the bitters adding the coup de grâce. Bitters? That's like drizzling Tabasco onto crème brûlée. After the effort that goes into making the stuff all that's needed is to ease the cork, pour carefully, and enjoy. You don't even need a corkscrew.

The polished brands, perfected over decades and blended across different vintages, different plots, and different grapes, offer seamless delight and palate comfort, while the quirky newcomers challenge the taste buds. The former is like an orchestra, where the whole is greater than the sum of the parts. The latter is akin to a soloist. The orchestra must not play in a done-it-all-before fashion, and the soloist must not be prone to attention-seeking histrionics. Currently, some of the "soloists" stand out because they are quirky, but quirk alone will not sustain consumer interest. Oddness doesn't always equate to quality.

Which do I prefer? When they are good I love them both. The big plus is that the tension, the jostling one-upmanship between big and small, allied to increasing competition from a host of other sparkling wines, is driving up quality. And that competition is getting more intense all the time. Champagne is well placed to respond—so long as it concentrates on upping quality in the bottle and not relying on self-anointed superiority.

There is more sparkling wine made elsewhere in France than is commonly supposed. "Crémant," meaning "creaming," is the generic descriptor, though cream is not the immediate impression when some of them hit the palate. Millions of bottles are produced and four of the best known are: Crémant d'Alsace, Crémant de Bourgogne, Crémant de Bordeaux, and Crémant de Loire. Each hails from a region better known for its still wines. Thus the Crémants taste a bit slack when it is known that superlative wines emanate from the same vineyards—it is hard not to look over your shoulder for a great Riesling when sipping a

Bubbling Syrah

Some people make bubbles simply for the fun of it, playing around in their cellars the way a chef might tootle about his kitchen in search of new creations. One such winemaker is Matthieu Barret of Domaine du Coulet in Cornas—traditional source of the sternest Syrahs. Cornas has adopted a friendlier demeanor recently, but bubbles?

At the end of an excellent tasting and impromptu lunch cooked over vine cuttings, Barrett blindsided his visitors by offering a taste of his sparkling Syrah, warning with a wink that it was explosive stuff. Opened at a safe distance and with the bottle pointed into a bucket, the wine surged out in a candy-pink geyser, a little of which was captured in the bucket. Polite sips were followed by polite comments and then it was back to the inky deep "proper" wines. With Cornas of such quality available, nobody lingered over the bubbles.

Crémant d'Alsace, or wish for a noble claret while drinking a Crémant de Bordeaux.

Despite a scattering of delightful examples, most Crémants are overshadowed by their senior siblings. They are drunk in their own regions and taste as if a local bigwig once decided it was beneath the region's dignity not to produce a sparkling wine. Too often the bubbles fail to mask underlying deficiencies. The crucial difference between Crémants and Champagne is that Champagne is the premier wine of its region; the still wines there are secondary. Known as Coteaux Champenois, until recently these were weak wines of little excitement; like watermarks, there was detail but no depth. Climate change has given them a shot of adrenaline and more are being produced, but for now they remain in curiosity corner.

* * *

Champagne is the daddy of all sparklers, but in recent years it has ceded much of the popular ground to Italy's lissome Prosecco, with Spain's Cava plodding along behind. That wild summary held water until recently, but now the ground is shifting and Cava has found its mojo, while Prosecco's fashion peak may have passed. Good Cava can now be counted as a rival to the other two and not the country bumpkin of old, to be tolerated when nothing else was available. As a result, and thanks to continuing improvements in other sparklers across the globe, the world beyond France bubbles brighter every year.

Prosecco was the beneficiary when 2008 went down a fiscal sinkhole. Champagne, the wine of overt celebration, was pushed to the margins; there was nothing to celebrate and nobody wanted to be overt. The less-ostentatious Prosecco chimed with the austere times post-2008. Being cheaper, it was less guilt-inducing and in a trice became the darling of the chattering classes. It was vinous virtue signaling as never seen before—mirrored by ordering tap water rather than bottled when dining out. Prosecco's image was well removed from the in-your-face brashness of much Champagne consumption, though it is difficult to account for its rampant success in recent times and Champagne hearties struggle to comprehend its appeal. To them, Prosecco is frail, tasting watery with some redeeming fruit notes. Yet that hasn't bothered legions of drinkers who quaff oceans of the stuff, attracted by the lightness the Champagne brigade dismisses.

Prosecco the wine used to be made from Prosecco the grape. Not anymore. In an effort to protect "brand Prosecco," the authorities renamed the grape "Glera" while retaining "Prosecco" as a designation of origin, to prevent wily operators outside the region from labeling any old bubbles "Prosecco." But Glera? Surely the land of Verdi and Puccini, of Donizetti and Rossini, could have come up with a less lumpen name. "Prosecco" rolls off the tongue; "Glera" tumbles out like an awkward child.

Meanwhile Franciacorta—Italy's finest sparkler—labors in the shadows, contenting itself with the approbation of a wine-loving coterie more taken with its gravitas than by Prosecco's facile beauty. It's a hardly surprising situation in a world that sees anodyne Blush Zinfandel getting the nod ahead of perky Chenin Blanc or transcendent Riesling. It's unlikely to change.

Where the grapes for many of the world's sparklers are well known, Cava's trio—Xarel-lo, Parellada, and Macabeo—remain obscure. And Cava itself has struggled to match Champagne's glam and Prosecco's cool. Where they were couth it was coarse and seemed content with its place in the doldrums, while the nobler bubbles challenged each other for supremacy. Cava comes mainly from northeast Spain but it can also be made in other parts of the country, not a circumstance likely to foster quality. Many wines with bubbles shelter under the Cava name, gaining some status while dragging down the reputation of the good stuff.

Change is afoot. Cava is sprucing up and now challenges for a place at the sparkling top table, courtesy of the launch in 2017 of

Sneer no More

In Italy's Emilia-Romagna is found the sparkler the world loves to sneer at: Lambrusco. Until recently I was happy to join in, based on my experience of gloopy examples with a ghastly cosmetic character. They were akin to liquid bubble gum and did a hatchet job on Lambrusco's reputation, similar to that visited on Beaujolais by *nouveau*. The real thing is a different animal: fleshy-purple in color, exuberantly fruity on the nose, and faintly bitter on the palate. It delivers a wake-up call for seen-it-all-before taste buds, lulled into languor by too many svelte, smooth, and always expensive wines. Only in Bologna can it be fully appreciated. The wine may travel but is best enjoyed in situ and left at that.

the *Cava de Paraje Calificado* and the foundation a year later of *Corpinnat*. The former imposes much stricter regulations on producers, specifying vineyard origin and controlling winemaking practices, while the latter is an association of traditional producers who have broken away from the Cava appellation. Included in their number are Gramona and Recaredo, both of whom have glittering reputations for long-aged sparkling wines that bear comparison with the best. After decades of drabness, Cava's socks are being pulled up.

There's hardly a wine-producing country that doesn't try its hand at bubbles, from Chile to New Zealand, South Africa to the United States, Brazil to Portugal. Most of the results are serviceable and fall into the same category as France's Crémants—"afterthought" wines that fill a need but don't compete with the headline-grabbing still wines. There are exceptions, but much palate punishment can be endured before they are found. Far better to seek out sparklers that are standard-bearers for their origins.

Two outposts from opposite ends of the globe are current favorites: England and Tasmania. Both might be designated as "new world," though doing so with England warrants explanation. It is geographically part of Europe but in other respects, politically and commercially, ol' Blighty seems determined not to be, as evidenced by the recent Brexit ballyhoo. While some will write doctoral theses to explain Brexit, wine lovers will rejoice in the increasing cohort of delicious, exuberantly fruity wines asserting themselves as standard-bearers for Shakespeare's "sceptered isle."

The best known is Nyetimber, for years a lone voice but now joined by others such as Wiston, a Nebuchadnezzar of which was used to launch the cruise liner *Britannia* in 2015. The Wiston rosé bounces on the palate, a strawberry-infused delight, to be poured at Glyndebourne and Henley picnics. Others include Bride Valley, Gusborne, and Hambledon. At present their cornucopia of flavors is less elaborate than the established Champagnes; it narrows toward singular fruit without the amplitude of true greatness. But as night follows day, toasty complexity will follow zippy fruit and, given the current rate of development, England will be a world leader by mid-century. Will Champagne be eclipsed? Probably not, for some Champagne houses have already backed the English horse by investing in vineyard land there. Such a move represents a closing of the circle, back to England where glass bottles were first made strong enough to withstand the pressure of all those bubbles.

Tasmania sings a different tune. Australian wine surfed a wave of success in the late twentieth century, largely reliant on strapping

Great Grape: Pinot Noir

Pinot Noir sits at the center of a whirlwind of cliché and superlative, a juxtaposition of pantomime villain and prima donna: from heavy-hoofed clown to Mozart's Queen of the Night. Regardless of numerous encounters with the former, the latter keeps wine lovers coming back for more. For eons, writers have ransacked their lexicons to try and capture Pinot's character: fickle as the Irish weather, bewitching as a temptress, enchanting as a nightingale. It frustrates and thrills and, when thrilling, it is incomparably brilliant. Most of that brilliance, and much of the frustration, is found in Burgundy's Côte d'Or, where the vines are tended with a manicurist's hand. The best results shimmer crimson in the glass, are scented on the nose and supreme on the palate—wines of transcendent beauty and allure that challenge winemakers worldwide to match them.

Until recently few managed to do so, their failure laid bare in their wines' shortcomings: jammy fruit, high alcohol, a coarse finish. Only the most dogged persisted, and persisted, so that today Pinots of real class are being made in Germany, South Africa, Austria, New Zealand, the United States, Australia, and closer to home in Alsace, where once it was an also-ran. Rejoice, Pinot-philes.

flavors and Crocodile Dundee chutzpah. Subtlety was not their calling card. What Tasmania could deliver—less holler, more harmony—was little valued. But it is in today's less-is-more world. Tasmania's cool climate is now her trump card, and the Burgundian pair—Chardonnay and Pinot Noir—are mimicking what they do in France: producing delicious still wines as solo players and equally delicious fizz when combining as a duo. At present they're getting less recognition than they deserve, probably because Australia is a well-established player; hence these sparklers do not have the same novelty value as their English cousins. Unless your memory stretches back to the time of Henry VIII, when a fair amount of wine was made in England, you will only associate that country with muggy beer; crisp sparkling wine to rival the best is a sit-up-and-take-note addition to her beverage roster.

England and Tasmania have come late to the sparkling party but looking askance when offered either is yesterday's response. For the average consumer they are still under the radar and if chanced upon are regarded with suspicion—a reaction that echoes the skepticism that greeted the early New Zealand wines. Everyone was taken aback by an agricultural product from there that wasn't kiwi or lamb—and legions of former skeptics now consume epic quantities of her Sauvignon Blanc.

Don't be one of the sparkling herd; strike out and try something new from England or Tasmania. We have yet to see the best of both; their industries are still in their toddler phase, and in time they will be Champagne's strongest challengers for supremacy at the quality end of the sparkling spectrum.

4

LEGACY WINES

We don't shun a valuable inheritance, or do we? The wine world has a legacy of treasures from an age long gone yet they are treated as outcasts. The great triumvirate of fortified wines—Port, Sherry, and Madeira—have been handed down across the generations and they are especially precious because, should they ever disappear, they will never come back.

Fortified wines are strengthened by the addition of raw brandy, a practice developed to help them withstand transport by sailing ship, where a voyage's length was dictated by wind, not will. At their destination they were found to be improved by the spirit. Wines that were previously coarse now glowed on the palate, where once they growled. However, if they didn't exist today nobody would invent them; bottling at the winery, refrigerated shipping, and precisely timetabled transportation make fortification unnecessary. The idea would not even occur to winemakers.

Tasting the Century

The "Vintages of the Century" tasting held in Porto in June 1999 saw a total of 271 Ports from 1900 to 1994 gathered for tasting by a host of experts from across the globe. It was superbly organized with an army of skilled staff on hand to ensure everything ran smoothly. The setting was the old stock exchange building, an appropriately grand backdrop for such an event.

At the end of the day many wines jostled for position of outright favorite, Graham's 1970 and Sandeman 1945 among them, but by common consent the Niepoort 1927 was finest of all. At seventy-two years old it was still a "big" wine, vigorous and concentrated, with the age only showing on the finish. My note from the time reads: "Dark walnut with a distinct nose of prunes. Seems to have everything in just the right degree. No component standing out. Fabulous balance and integration. Finely flavored with a certain lightness at the finish." I haven't tasted it since but live in hope.

Ponder a world where fortified wine is no more real than science fiction. Then some Douro Valley winemakers play around with a few barrels to make a new drink called "Port." Pleased with the results, they seek official recognition, but the authorities issue a blanket ban on such adulterated wines. The winemakers have hit the bureaucratic buffers, which is what would happen if somebody did happen to invent Port today. In the case of Sherry nobody would set out to make it because of the capital costs, with no prospect of a return from today's feeble prices. Only because the vast bodegas are already built and paid for is it viable and even then it struggles. Sherry sits in the paradoxical position where its reputation is peerless while its image is woeful. It is one of the world's greatest wines, with a reputation earned over centuries. And a reputation *is* earned, while an image can be bought off the shelf or be concocted by an influencer.

It is only because these relics utilize historically approved practices that their production is sanctioned today. If we fail to preserve them they will be gone forever, like any endangered species. Previous generations perfected them, yet today these gems are shunned by consumers. Legacy wines have been canceled. Gambling away an inheritance pales by comparison.

Port is often cast as the court jester of the wine world, but Port is resilient; it has weathered worse than ill-judged scorn and clumsy humor. Mention vintage Port and watch the tired caricature of the Port drinker totter onstage: the cussed colonel harrumphing from his leather armchair, the terror of his club, tolerated because it is hoped he will leave a fortune to his fellow members when he shuffles off. There's more to Port than that.

Port comes in many guises, ranging from the twin peaks of vintage and long-aged tawny down to humdrum offerings that deliver a hit of alcohol-laced sweetness without stirring the senses. At the base of the pyramid are ruby, white, and pink Ports. Ruby carries a faint echo of the sublime delights of a great vintage. It's a chink of flavor that engages enquiring palates, and it's worth investigating further.

Where ruby is a door opener, white Port is a dead end. It leads nowhere, and drinking it straight is a masochist's pleasure. Its one redeeming quality is that it makes a refreshing cocktail when mixed with tonic water and garnished with mint. This is best enjoyed lazing on a boat on the River Douro, Port's home, where the scenery is rugged and wild. Should you ever manage to visit, carve an hour from the schedule and find a boat. Arm yourself with the drinks as well as an extravagant supply of toasted almonds and set off at a dawdle. Sip and munch and revel in an array of flavors as dazzling as the Douro's summer sun: the nutty, creamy, salty tang of the almonds counterpoised by the prickly, fruity, sharp cool of the cocktail.

Hot Tongs

In response to an earlier announcement, every resident of the country house hotel crowded into the drawing room to see the "cabaret": me opening a bottle of Taylor's 1955 Port by using hot tongs. Port tongs are designed to open old bottles where the cork is so fragile it would crumble to smithereens if a corkscrew was used. The two semicircular ends clamp around the neck, but first they must be heated.

I did this by placing the end of the tongs into the log fire for fifteen minutes, while explaining the procedure to my audience. Then, moving swiftly and with a final check to see that cameras were at the ready, I clamped the red-hot metal around the neck for perhaps a minute, removed it, and immediately wiped the neck with a cloth dipped in iced water. With a faint click, the glass broke in a neat circle, just above the bottom of the cork, allowing the upper two inches of the bottle with cork inside to be removed easily. The wine was glorious. My Port tongs only see intermittent use, but the cabaret is always entertaining. Keep the youngsters well back though—their magnetic attraction to intense heat and broken glass could ruin an otherwise smooth performance.

Pink Port has no redeeming qualities, and drinking it is an exercise in palate abuse. It might be okay submerged in a garish cocktail, but apart from that there is little use for this lurid liquid. Perhaps it is luminous and glows in the dark? I have never checked. Best to move up the quality ladder.

Doing so brings ample reward, wines with something to say for themselves: Late Bottled Vintage, or LBV, and ten-year-old Tawny. Where vintage Port is bottled after two years, LBV has its maturation advanced by spending four years in cask before bottling. Tawny, on the other hand, is matured solely in cask and the age statement represents an average of the wines contained in the blend. This pair deliver real interest and offer the best value, but there is a "but." Both are satisfying up to a point and that is reached when their elders and betters are tasted.

Few wines can compare to a vintage Port from a great year, drunk at thirty years or older. It's satin laid on gold, commanding and comforting, and is only rivaled for supremacy by long-aged tawny, dated twenty or thirty years, sometimes forty. Thirty-year-old is the apogee, by forty the bones are beginning to show, not unpleasantly, but these oldsters can flex on the palate in a way the others don't. It is difficult to compare vintage and tawny. Vintage is all plump fruit and licorice, whereas tawny boasts a sterner snap of spice and toffee. Both can be supreme and should be served cool (not chilled, just cool).

Unfermented sugar gives Port its sweetness; the fermentation is stopped by the addition of grape spirit with a strength of about 80 percent alcohol. That's twice the strength of regular brandy or

Before or After?

The firewater spirit, of a much higher quality now than previously, is added when the fermentation is still bubbling away, and it stops the action like a train hitting the buffers. It's akin to electrocuting the yeast. The timing of the addition—whether before or after the fermenting wine has been drained off from the pips and skins—makes a palpable difference to the wine. If it is added with the skins still present, the alcohol extracts coarse, rough-hewn flavors alien to the satin caress of fine Port. Loosened by the hit of alcohol, gremlin tannins come forth, and rather spoil the party by their aggressive presence. Added after, with no solid matter remaining, the alcohol slips in under the plump fruit, underpinning it like the unseen foundations of a building, adding depth and breadth and a warm caress on the finish.

vodka; this is alcohol gone nuclear, and tasting it is like placing a hot ember on your tongue. It makes an incendiary hit, never to be forgotten. The alcohol kills the yeast, and the sugar is held captive by the spirit when it naturally wants to fuel further fermentation. Instead, it is preserved to provide the smooth topping for the firmer elements in the wine.

If any wine captures the essence of its origin it is Port, and a true understanding can only be gained by visiting the Douro Valley in northern Portugal. Doing so was once a chore, a trial to be endured by hardy wine merchants and others of that ilk. The beds were lumpy, and so was the food. Today, it is open to visitors like never before, beds are welcoming and food is satisfying, though masochists can still find relics of the past. It is best to explore the city of Porto first, preferably on foot, including a walk across the Dom Luís I bridge to Vila Nova de Gaia. This is where the Port "lodges"—vast warehouses of aging wine that double as museums and that are now open to visitors as never before—are located. Then take the train to the Douro. It clatters and bangs, and seems to spend as much time stationary as moving. Frustration threatens. That is forgotten once the valley is reached and the train threads its way between river and slope on the single line like a tightrope walker. At times you could almost reach out and touch the water. Describing this place in words is a challenge; every superlative falls short. Its primal beauty can only be hinted at, preparing the visitor to be stunned. It is harsh and captivating, with jaw-dropping vistas of terraced vineyards rising steeply from the riverbank. Because of its remoteness, the Douro engenders a sense of dislocation; you sink into it, losing connection to the outside world. It floods every corner of mind and imagination, and the torrent of images prompts sensory overload. A glass of Port eases the struggle to grasp it all.

In the words of Shakespeare's Sir John Falstaff, "If I had a thousand sons, the first human principle I would teach them should be, to forswear thin potations and to addict themselves to sack."

In this case, "sack" refers to "sherris sack" or Sherry in today's parlance. Leaving aside that linguistic obscurity, the entreaty to "addict" will raise the hackles of health fanatics, and Sir John's battle with the booze—he sank into bibulous disrepute—will be rallied to their cause. The upshot? His advice will not be used as a promotional slogan anytime soon, yet Sherry needs its staunch advocates as the twenty-first century enters its middle years if it is still to be around at its close.

Where Port is the stereotyped preserve of grumpy bores, Sherry's cast of caricatures is more genteel: maiden aunts, country

Close Relatives

It hardly needs stating that Sherry and Champagne are the finest aperitifs of all yet, despite the styles being markedly different, that is not the only thing the two wines have in common. They both spring from vines planted in remarkably similar chalky soil and the base wines gleaned from each are barely worthy of note. Both are then highly manipulated to yield the finished product. Huge effort is expended to craft the wines and, at their best, both represent the summit of the winemaker's art, for their character and style is derived from winery practices, not vineyard practices. Despite a recent move that sees the source of the grapes being given greater prominence this remains true for almost all Sherry and Champagne. Another similarity is that the big-name brands in each case offer reliable sources of quality. This brings us to the one glaring difference: price. Depending on the market a bottle of high-quality Champagne, say Pol Roger, will sell for about three times its Sherry equivalent, say Tio Pepe. Why the disparity? Popularity is the only explanation. If Champagne became unfashionable tomorrow its price would plummet. As with all business you charge what the market will bear.

vicars, and family solicitors, all far removed from the garrulous Falstaff. They sip their Sherry decorously, nostrils twitching at its dubious quality—wondering was it left over from last Christmas, or the one before. Those doornail-dead Christmas bottles usually contained sticky-sweet brown concoctions that utilized the sugar to mask all deficiencies. The anesthetic glow deadened Auntie's conversation, softened the vicar's news of bereavement, and combatted the solicitor's teasing out a fine point of law.

For much of the twentieth century, the Sherry party was the social gathering where everyone could meet and consume alcohol. The pub was a male preserve where beer was sluiced back as tall tales were exchanged, boredom held at bay by the bellyful of beer. Sherry was respectable and a modest measure could be stretched over an hour with no loss of decorum. Sherry party hosts seemed to vie with each other to find the vilest stuff to press on their guests, who took revenge in turn by finding something worse.

While in college I attended a number of these parties, usually held by clubs and societies to kick-start the year's social season, and as soon as the Sherry ran out it was off to the college bar for pints. But first we wrestled with the Sherry, using it to fuel stop-start conversations with senior club notables. At one such gathering I met a well-cut chap, grimacing as he sipped his drink. "I prefer Tio Pepe," he said, explaining that it was proper Sherry, dry as flint with a pale-bleached color and a "laser sharp" flavor. "You should try it," he said on departure.

When I took his advice, my palate was left reeling by a wine I thought shrill and searing, an experience I likened to giving up sugar in coffee. Without sweetness, the flavor stood exposed in sharp outline, and only minutes later I noticed how unsullied my palate felt. Lip-smacking freshness lingered, not a leaden resonance of sweetness. In time, as with coffee, I came to adore the lean flavor and today I count Fino Sherry and its cousin, Manzanilla, as two of my favorite wines.

The Sherry hierarchy is not as easy to grasp as Port's and that exacerbates its image problem. However, Sherry's trump card comes courtesy of the most lyrical quintet of names: Fino, Manzanilla, Amontillado, Palo Cortado, and Oloroso, each a style in its own right. Sherry marketing should make more use of them, while easing "Sherry" to the side where its dull image can fade away. Each should be championed, as I do when offering a glass of Manzanilla for instance, explaining that it is one of Spain's greatest wines. Elaborating, I'll tell my guest to expect a savory bite on the palate and not the fruity tang of more mainstream wines. And if some acorn-fed Iberico ham is served with it, even the most skeptical is won over.

Unlike Port, Sherry is fortified after fermentation, a practice that marks the essential difference between them. The base wine, made from the Palomino grape, is a blank canvas upon which the masters of the bodega work, conjuring flavors unimaginable at the outset. The wine is assessed and graded into two categories: lighter wines that will be gently fortified to 15 percent, and more robust ones that have their alcohol level pushed toward 20. The two basic

Great Grape: Palomino

Few grapes are as cloaked in anonymity as Palomino, yet it is the grape upon which nearly all Sherry is built. It is the unsung hero of the wine world, its light hidden beneath layers of biological and oxidative aging. Sherry gets most of its character from the production process and not its constituent grape, which, when vinified as a table wine, yields an ordinary wine of little impact. In this guise Palomino lacks vim or substance, a frail child in need of building up. That happens in the bodega where years of manipulation eventually see it emerge in a myriad of styles, each compelling in its own right. The complexity and depth of flavor of the best Sherry compare with the world's greatest wines, yet it remains hidden in plain sight; everybody has heard of it, few drink it.

styles—Fino and Oloroso—constitute the starting point for almost all sherries.

The wines are transferred to six-hundred-liter barrels, called butts, which are not filled completely, leaving the wine in contact with air, a practice that would ruin other wines. Nor is a barrel filled from empty with young wine. Instead, the youngster becomes part of the *solera* system, where fractional blending is used to ensure consistency of style year after year. A *solera* consists of perhaps a couple of hundred barrels arranged in tiers. Wine for bottling is drawn from the oldest tier and the deficit is made up from the next oldest and so on until the youngest tier is replenished with the new wine. No barrel is ever emptied; a well-managed *solera* can have a life span stretching over a century.

Finos and Olorosos age differently. A yeast called *flor* grows on the lower alcohol Fino, covering the surface and protecting it from oxidation. No *flor* grows on Oloroso, so it oxidizes, getting darker and more intensely flavored. If the *flor* on a Fino dies, the wine develops into an Amontillado, a style that combines elements of both and which can be the most complex of all Sherries. Another style, Palo Cortado, resists definition and is described obliquely as having the nose of an Amontillado and the palate of an Oloroso. Finally, Manzanilla is made like Fino but in the seaside town of Sanlúcar de Barrameda, which is said to give the wine a saline tang.

The Sherry cast list is completed by Pedro Ximénez, or PX, which takes its name from its constituent grape. After picking, the grapes are dried to concentrate the sugar and are fermented into a treacly wine that tastes of dates, prunes, and figs enriched with licorice. It is often poured over ice cream in place of hot chocolate sauce. PX is a layby on the Sherry highway and is the one style that is seldom referred to as Sherry. Thus, when I presented it at the end of a wine dinner as an unusual style of Sherry, a lady took issue and stated it was no such thing. As our spirited exchange drifted toward argument, I retreated with a vague platitude.

Forgotten Cousin

Created in 1773, Marsala will celebrate its 250th birthday in 2023, though until recently it looked like it might not see that milestone. For much of the last century it was known largely as a cooking ingredient, and up to twenty years ago it was almost impossible to buy anything other than drab stuff destined for the pot. Marsala was the brainchild of a wine merchant from Liverpool, John Woodhouse, who planted vineyards in the west of Sicily and who got his break when Admiral Horatio Nelson ordered huge quantities of it for his warships in place of the more usual rum.

For a period it enjoyed great popularity but then catastrophic decline set in, far greater than the mixed fortunes that befell the other "legacy" wines. Today, a flickering revival is taking place that might secure a niche place for Marsala in the vinous firmament but the glory days are over. It is made principally from the Grillo grape and comes in a range of styles. Some of the best do not qualify for the appellation because they are unfortified, but that technical impediment should not stop the curious palate from seeking them out. The search will be a little easier—and more rewarding—than it was until recently.

The Sherry bodegas are architectural wonders: high-roofed, white-painted, and stocked with thousands of blackened oak barrels. Outside, the summer sun dazzles off the walls while the interior is dim and hushed and remarkably cool. Ranks of barrels recede into the darkness and the sandy floor softens sound. Time could stand still here. A pleasant whiff of wine and wood and earth tantalizes the nostrils. The appetite stirs.

It is time to taste. The cellar master stands ready with his *venencia*, a slim metal cup with a long flexible handle. He plunges it into a barrel to extract a sample and then pours it masterfully from a height into your *copita*. The stream of liquid dances and sparkles, in contrast to his insouciance. The performance is executed with a ringmaster's flourish—participate in it and you will be hooked for life. No matter what you taste, from delicate Fino to rugged Oloroso, a comforting savor lingers on the palate, a baseline flavor, like a familiar jawline in siblings of otherwise different appearance.

The best Sherries are thrilling, but the fly in the ointment is the ocean of junk churned out to satisfy commercial demand. As Port has its howlers, so too does Sherry, with Pale Cream standing accused. It's a timid wine that creeps over the palate inducing "taste bud torpor." Such dross should not blind us to the magic that is great Sherry. The transformation effected in the bodegas—from forgettable to miraculously flavorsome—is nothing short of sorcery. Celebrate sack and sorcery.

"How long has that been there?" I asked a friend, pointing to a half-finished bottle of Madeira sitting on a windowsill, exposed to the sun. She guessed a couple of months, adding it was time she got rid of it, as it would obviously be undrinkable. Before she took such drastic action, I suggested a sip and was rewarded with a delicious drop that somersaulted across the palate. It was everything that good Madeira should be, caramelized fruit and burnt toffee leavened by a ping of acidity and topped with extraordinary length. She stretched the remainder across another few months.

Six centuries ago the Portuguese landed on the densely forested island of Madeira, six hundred miles southwest of Lisbon, and set fire to it. The fires burned for years, coating the land in a layer of fertile ash. Madeira means "wood" in Portuguese, and the island is little more than a spike of land poking through the ocean, clad with sheer cliffs and rising to six thousand feet. Terraced vineyards cover its lower slopes, planted with the grapes that name the wines in ascending level of sweetness: Sercial, Verdelho, Bual (or Boal), and Malmsey (or Malvasia).

When wind power was the only means of crossing the oceans, Madeira's position on the routes to America, Africa, and Asia made it a perfect victualing station. Barrels of wine fortified by a couple of buckets of spirit were loaded aboard the sailing ships, to share space in the hold with the slop of bilge water, while also acting as ballast. Paradoxically, the long voyages in all weathers—and the stifling heat of equatorial waters—markedly improved the wine. As a result, Madeira enjoyed huge popularity in eighteenth-century America, particularly Savannah, Georgia, where it is still revered. Madeira was used to toast the Declaration of Independence in 1776 and the United States remains one of the best places to find a rare old bottle. Bern's Steakhouse in Tampa, Florida, which has one of the greatest wine lists in the world, is a good starting point.

In time, canny producers took to mimicking the trial-by-sea-voyage, warming the wine for a couple of months in *estufas*, storage spaces or tanks heated to over 104°F. The practice is still employed for commercial Madeiras, while the best wines are given an extended maturation in six-hundred-liter casks stored in the eaves of the producers' lodges, well warmed by the sun. The wine eventually emerges racy and sharp with a spine of acidity like a lode of precious metal. A top-quality Madeira might spend twenty years in the barrel before bottling is even considered and then it might not occur for another few decades.

Thus forged in a process akin to tempering steel, Madeira is the world's most extraordinary wine, impervious to hardship or abuse. It is the wine that should not be. Subjected to every manner of maltreatment, it bounces back strengthened and vigorous. Warmth and oxygen, the enemies of wine, are essential to it. Madeira converts them to staunch allies and, far from damaging the wine, they reinforce it.

While open bottles keep for months, unopened ones can last centuries. At a dinner in 2018, a 1900 Boal, Manuel de Sousa Herdeiros, was served and as a dozen glasses were poured at a side table the room was filled with a vivid aroma of caramelized fruit.

Cool It

Neglect from those who are indifferent to fortified wines is one thing, but a greater crime against them is regularly committed by those who love them and who should know better. With the exception of Fino and Manzanilla, fortifieds are almost always served too warm. To show at their best, they should be served cool rather than chilled. Something in the range of 52° to 57°F will serve for the majority, and this is well below room temperature. If they are mistakenly served too cold, a small measure in a glass will quickly come up to the correct temperature, at which they will have a precise, well-defined flavor.

If, for instance, a vintage Port is served too warm the alcohol and sweetness, previously woven together, disengage and go their separate ways, leaving a wine with a "blurred" flavor. The correct temperature acts like a boundary on the flavor, holding it in shape, giving it impetus. It takes little effort to get things right, but doing so brings rich reward. Cool it.

No other wine could have scented the room so. The flavor was more forceful, led by a punch of singed fruit, like the crust of a Christmas pudding scarred by flaming brandy. But it was still vital, with none of the disintegration of an over-aged wine, where the flavor is like a jumble of broken-down furniture in a junk room.

Of the fortified triumvirate, Madeira is the least popular, but its day will surely come, for its potent acidity chimes with the current zeitgeist for fresher wines. The acidity sets it apart, a rapier thrust that grants mythical powers of longevity. To categorize Madeira simply as "fortified" barely captures its resilience; this wine is as close to indestructible as any living thing can be. It shrugs off the decades as they turn to centuries, remaining vital and engaging when even the greatest of Ports are heading for the grave. Such rare attributes deserve a wider audience.

A great Madeira needn't be a pensioner to sing like a diva; because of its ageless quality it can drink superbly over many decades, from youth to antiquity. A Barbeito Malvasia 2000 Single Cask 39 "a+e" I enjoyed in 2013 was vividly memorable, the "a+e" referring to the two warehouses where the wine aged. It started out in "a" and was then moved to the cooler "e" to slow the process and give better definition to the acidity. An ethereal perfume presaged a palate of fantastic depth with lovely texture and not too much weight. There was citrus, spice, and nuts; it was opulent yet invigorating with no lushness and paired beautifully with the bitter chocolate tart served alongside. It was a reminder that for thrilling intensity of flavor, there is nothing to beat Madeira. Nothing.

5

ACROSS THE POND

Frank Sinatra crooned that he wanted to be part of "New York, New York," echoing the hopes of generations across the globe. Mention of the great metropolis prompts dewy-eyed reveries about visits once made and others anticipated. New York has it all, yet the public imagination seldom strays beyond Manhattan. If it did it might drift some 250 miles upstate to the Finger Lakes, where rural tranquillity contrasts the city's hustle, especially late in the year when the forested hills blaze with autumn's yellows and oranges and reds. Come winter, the temperature can drop low enough to kill the vines, but the region can cope thanks to the work of the man considered father of the modern Finger Lakes wine industry: Dr. Konstantin Frank.

He arrived in the region in 1951, though the first vines had been planted in 1829 when the Reverend William Bostwick laid out a vineyard in his rectory garden in Hammondsport, followed in 1853 by Andrew Reisinger, who planted a further two acres. As the century progressed, a fledgling wine industry developed, given impetus in

1873 when a Finger Lakes "Great Western Champagne" won a gold medal at the Vienna Exposition. These original plantings were of American *Vitis labrusca* vines such as Catawba and Isabella, hardy but humble, able to withstand the cold but incapable of matching the nobler *Vitis vinifera* vines for quality. Frank, a Ukrainian native and unfazed by cold winters, was convinced that *vinifera* vines, if grafted onto tough rootstocks and carefully cultivated, would withstand winter in the Finger Lakes.

Through the 1950s he worked to verify his conviction and win over the skeptical locals. A severe frost in February 1957 caused huge damage across the region, while Frank's grafted vines came through with only minor damage; the doubters were convinced. He had most success with Riesling, now the region's standard-bearer, the best examples of which exhibit a delicate precision that calls to mind the filigree delights of Germany's Mosel Valley.

The Finger Lakes' autumn splendor renders it one of the most beautiful wine regions, cherished by locals but largely unheard of by the global wine community. Its name derives from the series of roughly parallel, pencil-slim lakes that can be seen as the fingers

Foxy

"Truly awful," is how I once described the flavor of wines made from *labrusca* grapes. The usual descriptor is "foxy" but that barely captures the heavy, feral odor they exude. They fume in the glass, churning out a noxious fog that lingers in the nasal passages. The palate follows the nose, boasting flavors that lie beyond the normal descriptive lexicon, with its pleasant litany of fruits and spices and herbs. There is nothing pleasant about these concoctions. "Skunky" is another descriptor, and it captures well the ghastly whiff of wild garlic running through the fruit. As with the smell, the flavor lingers, only being banished by a cold beer. Avoid.

of a hand. The beauty belies the force of nature that created them, as retreating glaciers gouged the Earth's surface, to a depth of over six hundred feet in what is now Seneca Lake. The lakes run roughly north-south, with woodland and vineyards between them. The bodies of water mitigate winter's cold, giving off summer's accumulated warmth, and without their influence not even Frank's viticultural wizardry could have made vine growing viable here.

Whereas the successful cultivation of *vinifera* vines in the Finger Lakes was the turning point in the region's history, no such drama attended the development of the quality-oriented wine industry on Long Island, a two-hour drive from Manhattan. The island's wine hub is Sag Harbor, an old whaling village whose name derives from the word *sagaponack*, meaning "land of the ground nuts." This is old America, dating back to before the 1776 revolution, and a genteel air pervades, at variance with the brashness of Manhattan. It contrasts also with the rough and tumble life when whaling was at its peak and the town was filled with hard-living sailors. Strolling the streets of Sag Harbor today, it is hard to imagine those days—one can't picture the whalers mulling the merits of a Cabernet Sauvignon or debating the age-ability of a Merlot.

Long Island forks north and south as it reaches into the Atlantic, and the frigid challenges of the Finger Lakes are unknown there, the ocean softening the winter-summer divide. The more sheltered northern fork is home to the majority of the wineries, while the southern serves as a playground for Manhattan socialites in the Hamptons. The maritime climate, along with the relatively flat vineyards, calls to mind Bordeaux, and the older region's grapes feature prominently: Cabernet Sauvignon and Cabernet Franc, along with their brother-in-arms, Merlot. In style, the Long Island reds echo Bordeaux's harmonious clarets, suffused with lingering but never coarse flavors. Long Island's "clarets" have yet to scale those vinous heights, but I would mention the best of them in the same breath as Bordeaux.

The best include a bottle of Paumanok, Cabernet Sauvignon 1995, Tuthills Lane Vineyard, carried home from Long Island after a Thanksgiving stay at The American Hotel in 1998. I finally opened it in 2020, convinced it would be dead, a feeling compounded when the corkscrew made dust of the cork. As I filtered the wine through a fine sieve, the glowing ruby-garnet color gave a glimmer of hope. Then came a waft of fruit, savory not perfumed, like an assuring pat on the back. The flavor was beautifully harmonious, its constituents woven seamlessly by a quarter-century so that trying to identify each was futile; far better to enjoy the whole than to burrow in search of components.

The American Hotel

Sag Harbor's American Hotel is famed for a wine list in which oenophiles happily lose themselves as their minds travel the world—from Bordeaux to Burgundy, California to Châteauneuf-du-Pape, the Loire to Long Island. Settling on a choice is a pleasant agony.

The hotel dates from 1846, when the whaling industry was at its peak, and stands on the site of an older inn where British troops were billeted during the revolutionary war. For much of the twentieth century its fortunes declined, exacerbated by war, Prohibition, and worldwide depression until by the 1970s the building was a near wreck. Its reincarnation began under the stewardship of Ted Conklin, who bought the premises in 1972 and set about undoing decades of neglect, including the bucket-by-bucket removal of tons of coal ash dumped in the cellar. Scale is everything at the small hotel; this is no soulless hostelry. Guests are welcomed, not processed; their enjoyment takes precedence over production-line efficiency. Then it is time to peruse the wine list.

The winemaker's blurb on the label—too often a vehicle for lunatic claims no wine could meet—for once rang true: "The Tuthills Lane Block from which this wine is made, was planted in 1982 and is our oldest. The grapes that the block yielded in 1995 were so ripe that we fermented them separately creating the finest quality red wine produced at Paumanok so far... Only 75 cases were made." Would that I had one of those.

I was lucky with the Paumanok, where age added luster. Too often, however, a wine's flavor is "front loaded," designed to dazzle in youth with no thought of a future beyond the immediate. In time the impressive musculature fades, leaving a wine coarse in spots and limp in others, with all the structure of a semi-deflated balloon. They croak rather than sing.

Such was the case with a Ravenswood Cooke Zinfandel 1996, which I carried home from California in 1998, when air travelers could still stash bottles in hand luggage. Upon discovering this gem far from home, the temptation was overwhelming, so I crammed in six bottles along with three others, turning my carry-on bag into a ball and chain in the days before wheeled luggage. When I was boarding my flight, the lady at the gate asked about carry-on luggage. "Only this," I assured her—as the strap gouged my shoulder. On board, when the flight attendant offered to stow my bag, I declined and suggested she stand back to allow me to swing it overhead like an Olympic hammer thrower.

It wasn't worth it, as I recorded twenty years later: "As I left the Ravenswood winery ... I put my money where my mouth was and bought six bottles of this wine. Five were drunk in relative youth and they lived up to the winery's faux Latin motto, *Nullum Vinum Flaccidum*, which roughly translates as "No Wimpy Wines." Would their twenty-year-old brother do the same? Not quite. The muscle and flesh of youth had yielded to bone and sinew, in the form of faded fruit, sawdust tannins, and a hot ping of alcohol on the finish. I'm glad I kept one bottle, but only one."

Despite early enthusiasm, I quickly tired of Californian Zinfandels and Cabernets that walloped the palate with look-at-me flavors. Ditto plush Chardonnays that slid like serpents, the sting coming when the hidden hit of alcohol struck home. Flavors became more and more concentrated, with more and more new oak barrels used to bolster them further. Making the loudest wine of all seemed to be the aim. Napa Valley Cabernet Sauvignon, a world-class wine in the 1970s and '80s, went pear-shaped as the century drew to a close, losing its figure in spectacular fashion, to the point where it was almost impossible to discern wine flavors among the concentration

Great Grape: Cabernet Sauvignon

The Médoc and Graves regions of Bordeaux chose well when deciding to build their reputations on Cabernet Sauvignon. Cabernet is reserved and composed, fruity but dry, never flashy or obvious; "charming" and "Cabernet" are not bedfellows. Instead of flamboyance, it possesses an architectural quality that sets it apart. Cabernet's identifier is blackcurrant. Blackcurrant supported by a backbone of tannin with added facets of cedar and a mild herbaceous whiff leans toward austerity, not opulence, and hence blends well with fleshy Merlot. Cabernet provides the structure, Merlot the soft furnishings.

Cabernet's upright personality remains immutable no matter where it is grown and it produces impressive wines in Italy, Australia, South Africa, South America, and elsewhere, yet its non-French heartland is California's Napa Valley. There, it is currently in recovery from the gross years when it was vinified into carthorse wines rather than the Thoroughbreds of earlier days. Today, wines of grace and elegance are prized once more, and palates previously mugged by grotesque flavors can relish the sweet savor and elegant reserve of proper Cabernet once more. King Cab? Indubitably.

and raging oak. The saddest case was the Robert Mondavi Cabernet Sauvignon Reserve, once a Napa standard-bearer, that foolishly followed the herd down the path of bloat.

Through all this one light shone, unwavering in its pursuit of excellence: Ridge. Amid the heat and dust generated by publicity-minded wineries, Ridge, in the Santa Cruz Mountains region, got on with its business, crafting wines of rare balance and purity and then letting them do the talking. Its reputation was honed courtesy of diligent labor over decades; spin and hyperbole played no part in it.

A trio of the flagship Montebello Cabernet tasted in 2012 confirmed Ridge's enduring quality. The vintages were 1995, 1997, and 1999 and the word "fresh" dominated my notes: "so fresh," "wonderfully fresh," "delicious freshness," and so on. Shouldn't all wine be fresh? Yes, it should leave the palate clean and ready for another sip or another mouthful of food, not reeling from brutish flavors. The 1995 was an essay in balance, gently fruited with a modest tingle of acidity, with some aged frailty on the finish; the 1997 was more vigorous with a satisfying texture and good length; the 1999 was the best, still with something in reserve, as evidenced by the mild streak of austerity among the abundant fruit. All were lovely.

California is the hub of the American wine industry, source of storied wines that deep-pocketed collectors fall over themselves to acquire, as well as the most basic plonk, churned out by the ocean-load. It is also home to the film industry, and California has conquered the world thanks to that. After all, film exerts huge influence over how we live: what clothes we wear, what songs we sing. And when film intersects with wine it dictates what we drink—and what we don't.

It did this most memorably in 2004 when *Sideways*, the adaptation of the eponymous book by Rex Pickett, hit the big screen and radically changed the fortunes of Pinot Noir and Merlot. The central character, Miles, was a "Pinot-phile" beguiled by its sensual beauty: "Its flavors, they are just the most haunting and brilliant and thrilling and subtle and ancient on the planet." He did not feel the same way about Merlot: "If anybody orders Merlot I am leaving. I am not drinking any f*****g Merlot." The film's release was followed by a planting frenzy for Pinot—and a stunning decline in Merlot's popularity.

Despite his trenchant opinions, Miles was a loser, at odds with the world and himself, sustained by dreams of success as an author. The film follows him and his soon-to-be-married friend, Jack, on a prenuptial stag week meander through the vineyards of Santa Barbara. The week progresses haphazardly, and mayhem ensues when Miles explodes in frustration on a winery visit and attempts

Great Grape: Merlot

Despite the pasting meted out to it by *Sideways*, and the resulting plummet in popularity, Merlot has deservedly survived. It might not have the cachet of its regular blending partner, Cabernet Sauvignon, but many Cabernet-based wines would be the poorer for the lack of it. Merlot is widely planted, a grape of easy virtue, adding flesh and sweet fruit to blends, often plumping out Cabernet Sauvignon's more severe features to yield wines of harmony and balance. In this respect it occupies a similar position in the firmament to Grenache, loyal support, appreciated for what it brings to a blend but seldom accorded standalone status.

Except, that is, in Pomerol and Saint-Emilion, where Merlot is the senior grape to the Cabernet cousins, Sauvignon and Franc. Here, the wines are sumptuous and verge on opulence, and the best are some of the most sought-after in the world, diva darlings of auction houses from Hong Kong to New York. It is hard to make any connection between these princely wonders and the drab concoction that is bargain basement Merlot. Is Merlot deserving of Miles's scorn? Yes and no.

to drink the contents of the spittoon. Through it all his love of Pinot runs constant, at odds with the unfolding chaos, and thanks to his championing of it the siren call of the temptress grape echoed around the wine world and beyond its cloistered confines too.

Years previously, I experienced a less chaotic wine-film intersection when visiting the Niebaum-Coppola winery in Napa Valley, founded in 1879 by Gustave Niebaum, a Finnish fur trader, and owned by Francis Ford Coppola of *Godfather* and *Apocalypse Now* fame. The tasting went to script: sample, spit and scribble, sample, spit and scribble . . . interspersed with the usual question and answer between visitor and host. Nothing out of the ordinary.

The upstairs visitor center was a different matter, a wonderland accessed by an ornate staircase crafted from exotic hardwoods from Belize: black poisonwood, grendillo, and jobillo. It was pure theater, with dramatically styled display cases whose angled lines mimicked a perspective drawing—all far removed from the normal ranks of oak barrels and stainless steel tanks that enthrall on first encounter only. Far removed too from the usual hawking of branded baseball caps, T-shirts, aprons, corkscrews, and other accessories.

It was like stepping onto a multiple-film set: here the surfboard from *Apocalypse Now*, there the desk from the *Godfather* and, in a corner cabinet, a clutch of Oscars to drool over. According to my host it took, "Time, vision, and money, mostly money," to create the spectacle—on a scale to match Coppola's films. The wines were impressive, though not in the same league as the celluloid masterpieces. I left a cigar for the legendary director—a Romeo y Julieta "Churchill" in a tube—at reception. I'm not sure he ever got it.

Theater of another sort awaited at Opus One, temple to wine and joint venture between Robert Mondavi of California and Baron Philippe de Rothschild of Bordeaux, whose silhouettes were etched onto the stainless steel tanks. The building's stark superstructure evoked a Bond villain's lair. Inside the hush was reverential and the chairs in reception were as uncomfortable as they were eye-catching. The inch-perfect building was matched by the wine, polished and suave, like a sun-bedded Las Vegas crooner. A small fault would have added attraction to the sterile beauty. The Bentley in British Racing Green parked outside completed the rich-man's-plaything picture.

In scale and reputation, Napa Valley calls to mind Burgundy's Côte d'Or, diminutive yet exalted. It is about thirty miles from north to south, never widens beyond five miles, and is the jewel in California's vinous crown. Its proximity to San Francisco makes it a mecca for visitors. Finding your way around is a doddle thanks to the main road, Highway 29, that runs through the valley like a spine, unlike the twisting labyrinth of roads and tracks in many old world regions. Napa could be driven in an hour, though languid exploration is more rewarding, detouring left and right to visit a roll call of famous names.

Still the most recognized is Mondavi, even if the luster has faded since its heyday when it was a byword for excellence and innovation. Founded by Robert Mondavi in 1966, it was the name that put Napa on the map. Mondavi was a commanding presence despite his small stature and high-pitched voice, and quality was still paramount in the 1990s. Having been an advocate for stainless steel fermentation tanks, Mondavi switched back to wooden vats, built in France, disassembled, shipped, and then reassembled *in situ* by a team of coopers flown from France.

A huge replanting program was also undertaken, though this was forced on him and many others by an outbreak of phylloxera, when a rootstock once thought impervious to the louse succumbed. Seeing opportunity in adversity, Mondavi chose to plant the vines at a much higher density than previously, necessitating the importation and use of *enjambeur* tractors to straddle the close-spaced rows. At the time there were only eight in the United States, five of which were owned by Mondavi at a cost of $120,000 each.

Phylloxera is the baleful tie that binds the European and North American wine industries. It is a tiny aphid that feeds on a vine's roots, eventually destroying it. In the nineteenth century American vines were impervious to phylloxera, having developed resistance

Hands across the Atlantic

Franco-American wine links abound. Nearly ninety years ago US banker Clarence Dillon bought the celebrated Château Haut-Brion in Bordeaux, and his descendants still own it, as well as its next-door neighbor, La Mission Haut-Brion. Another American banker, Robert Wilmers, bought Château Haut-Bailly in 1998, also in Bordeaux, leaving the previous owners in complete charge of day-to-day affairs.

In Burgundy, Domaine Dujac is the creation of Jacques Seysses and his Californian wife, Rosalind, and today their eldest son Jeremy takes charge of winemaking with his American wife Diana, who also makes wine at her family's Snowden Vineyards in Napa Valley. Aubert de Villaine, coproprieter of Domaine de la Romanée-Conti, is married to Californian Pamela and together with her cousin's family makes Hyde de Villaine wines, also in Napa. The Drouhin family, headquartered in Beaune, established an outpost in Oregon in 1987. Recent years have seen huge US investment in Burgundy: Silicon Valley entrepreneur Michael Baum bought Château de Pommard in 2014 while, two years later, Stanley Kronke, owner of cult wine Screaming Eagle, bought Bonneau du Martray in Pernand-Vergelesses. Investment went in the other direction in 2021 when Faiveley bought a stake in Sonoma's Williams Selyem.

Finally, a high-profile matrimonial link was joined in 2009 when Jean-Charles Boisset, scion of the wine business founded by his father, married Gina Gallo of E&J Gallo. Appropriately, they live in Robert Mondavi's old house, Mondavi having been one half of the Opus One joint venture.

through prolonged exposure, but when it crossed the Atlantic in the 1860s, the European vines fell prey to it and across the continent whole regions were laid waste. It was silent, barely visible, and deadly. An old vine, with a massive, gnarled trunk carrying the scars of decades looks rugged as granite, yet such outward strength was no defense against phylloxera. It was as unstoppable as an incoming tide and remains a worldwide threat today.

Phylloxera was the great leveler; it visited destruction on vineyards and despair on winemakers without favor, from humblest to most exalted. The only difference was that many lowly vineyards disappeared for good, while the exalted staggered on until even they succumbed. Ironically, salvation came from America—eventually—but only after every manner of scheme, madcap and otherwise, was tried to combat the tiny beast. One treatment—the injection of carbon bisulphide into the soil—was clumsily effective but impractical on a large scale. It also killed much else besides the louse, sometimes even the vine, as well as being injurious to the workers applying it when it could catch fire. Others were children of desperation, hare-brained notions fed by wishful thinking, such as the suggestion that a live toad buried under each vine would ward off the predator. It is easy to deride such efforts, but that ignores the reality of helpless peasant farmers seeing their livelihoods destroyed. Clear thinking is rare in such circumstances.

The solution was grafting, splicing European vines onto American roots, thus creating a historical interdependence largely invisible today. It worked, though it couldn't undo the decades of devastation. Problem and solution both came from America, laying low the European wine industry before saving it, transformed but viable again. Progress was slow as the vineyards of Europe replanted on a vast scale; Burgundy's Romanée-Conti, the most prized vineyard in the world, only accepted defeat in 1945 when the ravaged vines were grubbed up and replaced with grafted ones. It is barely an exaggeration to say that were it not for America the prized wines of Europe would not exist today. Phylloxera remains the most destructive enemy the wine industry has ever grappled with, though some are suggesting climate change may usurp it in time.

* * *

Having almost destroyed the European wine industry, America turned on its own, when legislators came up with a suicide scheme of rare ingenuity—Prohibition. If ever a piece of legislation spawned unintended consequences it was the Volstead Act, which added the 18th Amendment to the United States' constitution and came into force in January 1920. It marked the end of a long campaign by the temperance movement against the evils of alcohol, dating back to Maine going dry in 1851. After that, state followed state and over thirty were dry by the advent of World War I. Worthy motives begat clumsy methods, as a legion of shortsighted do-gooders got their way, sowing the wind with no thought for the whirlwind. They cannot have guessed what lay ahead.

Rather than saving the country from the scourge of alcohol, Prohibition did the opposite, while facilitating the rise of organized crime. Cynicism polluted public life; lawbreaking by otherwise law-abiding citizens was validated. Official America infected itself with a hydra-headed monster that has never been properly tamed. Though it lasted little more than a dozen years, its effects far outlived the 1933 repeal of the act. This was Prohibition's enduring bequest to the nation.

An unexpected consequence was the significant increase in grape production despite the ban on winemaking, this paradox explained by a huge increase in at-home winemaking. Thanks to a nebulous clause allowing for the production of up to two hundred gallons of nonintoxicating fruit juices for home consumption— non*intoxicating*, not nonalcoholic—a massive home-based wine industry blossomed overnight. Extraordinary quantities of grapes were required to feed it, and the growers in California responded with alacrity.

Freight trains of grapes crossed the continent to the big cities of the East, and demand was not for noble varieties but for humble grapes, ones able to survive the rough and tumble of a long train journey followed by unskilled winemaking. Alicante Bouschet, with hide-tough skin and dark juice, was the favored variety, endowed with such heft that the skins and pips could be flogged repeatedly to make subsequent batches of "wine" by adding water, sugar, and yeast, like reusing a tea bag, wringing ever more pallid concoctions from it. Quantity was never so valued over quality as during Prohibition. Shipping and at-home winemaking were further facilitated by the production of grape concentrate and dried grape "bricks"; more could be shipped and they were easier for the bathtub winemaker to handle. No instructions on how to turn them into wine could be given, but a po-faced warning not to add yeast for fear of initiating fermentation was included. Thus cautioned, the novice could proceed as he saw fit.

Adversity spawned ingenuity on a grand scale and gave rise to a slew of schemes to circumvent the law while not quite breaking it. In some cases the legislation aided its own flouting: doctors could prescribe alcohol for patients troubled by a persistent thirst, and production of sacramental wine for religious purposes was also permitted. Unhealthy yet devoted was the new norm for countless Americans, as phantom illnesses proliferated and religious observance soared.

Amusing as these tales are, they should not disguise the damage done to the regular wine producers forced out of business by the blinkered legislation. By the time things got going again, the United States was in the grip of recession and tastes had coarsened, corrupted by the lurid cocktails of the roaring '20s, often highly sweetened to obliterate all trace of the paint-stripper spirits used in them. Blunt flavors were in, subtlety out, and it remained that way almost until the 1960s when the wine industry flickered back to life.

Since then, numerous other states have emerged from California's shadow, notably Oregon and Washington and, more recently, Virginia. The first two lie respectively north, and farther north again, of California. Oregon's place on the world wine map has been won by classy Pinot Noirs, no matter what other varieties are planted there. The state produces little more than 1 percent of the nation's wine but, just as Burgundy's Côte d'Or enjoys a renown far

Sam's

Sam's Wines & Spirits in Chicago was billed as "The World's Wine Superstore," an idle boast, I thought, until I checked it out on a visit back in 2001. In appearance the premises at 1720 North Marcey Street were unremarkable, barely giving a hint of the excitement that lay inside. But once inside, it was as though I'd entered an Aladdin's cave of wine, by some distance the finest I have ever seen. An afternoon passed in a pleasant reverie as I perused aisle after aisle of vinous treasure; this was far more than a shop, this was a whole world of wine. No matter what I asked for, no matter how obtuse or arcane, an assistant pointed me straight to it.

Sadly, Sam's is gone, outpaced by the competition, and so is the Aldo Conterno, Barolo 1996 I bought that day. But I still have the bottle of Domaine Dugat-Py, Charmes-Chambertin 1998, the price sticker reminding me that I paid $120 for it. For now it remains safely tucked away, but I plan to pull the cork on its twenty-fifth birthday and raise a glass to Sam's.

in excess of its scale, so too does Oregon's Willamette Valley, source of Pinots that sit somewhere between Burgundy's reserve and the exuberance of other new world examples.

Washington is second only to California as the United States' largest wine producer and that, together with the dominance of Cabernet Sauvignon, cast it as Bordeaux to Oregon's Burgundy. Though the majority of vineyards require irrigation to be viable, in Walla Walla Valley, which straddles the border with Oregon, some vineyards are sustained by rainfall alone.

Way east of this pair, Virginia's wine country lies within an easy drive of the nation's capital, Washington D.C., and vine growing has had a long, if troubled history there. America's third president, Thomas Jefferson, having visited France shortly before the Revolution, sought to establish a wine industry in the state to combat the populace's fondness for strong spirits of dubious quality. Importing vines from Europe he tried and failed, tried and failed, unaware that his efforts, and those of others he encouraged in this crusade, were doomed by phylloxera that gobbled the newcomers' roots while leaving the native vines untouched.

Jefferson's story links the wine worlds, old and new, and weaves together the twin threads of America's varied experience with wine: Prohibition and phylloxera. He was no prohibitionist and sought a sensible solution to problems of over-consumption, unlike his successors who were driven by zeal, not reason. Today, the Virginian wine industry he strove to create is vibrant and hangs its hat on the scented Viognier grape. While that would satisfy him, what would surely surprise him is the fact that every state, even frigid Alaska and ocean-lapped Hawaii, produce wine of some sort. It surprises me.

6

DOWN UNDER

The winemaker glanced at me: "You don't want to taste my wines?" I shook my head. "But you'd kill for a cold beer?" "Yes!" I whooped. It was late afternoon at the St Hallett winery in Australia's Barossa Valley and a day spent tasting concentrated red wines had left palate and jowls battered. A Coopers Pale Ale was pressed into my hand, and its bitter-cold prickle jolted the taste buds back to life. We sat and talked about the wines, and thanks to that winemaker's clever thinking this is the only visit from that busy day twenty-five years ago I clearly remember. As they say in Australia: "It takes a lot of beer to make good wine."

Crimson? Vermillion? Scarlet? The flame red fuselage of the Qantas 747 Jumbo *Wunala Dreaming* turned heads and started people talking. Multicolored kangaroos frolicked on that red background in vivid patterns, looking like they might come to life once airborne. Sitting on the apron at Heathrow the plane was unmissable, waiting to take me on my first visit to Australia. It was bold and brash, like her wines and the characters who made them.

In the 1990s Australia was on a roll, as a new generation of drinkers discovered that wine didn't have to consist of swingeing acid, brusque tannin, and the like. Fruit could be ripe and succulent and garnished by oak, the latter adding a dollop of vanilla and a textural quality that smoothed any bumps in the flavor. Such wines became totems for Australia as they bounded onto the consumer radar. Brash and forthright, unconcerned about the rule book or centuries of tradition, keen to experiment and innovate, the winemakers cantered around the globe, eyeballing any who dared suggest their wines weren't up to the mark. Above all, it was their ability to produce immediately likeable and consistently dependable wines that won wine drinkers over in droves.

Common to many of them was an impression of sweetness, not a saccharine hit, more a softness that made the wines comforting rather than challenging. There was plenty of heft, but thanks to the "oak cloak" there was no hardship in drinking them and crucially they could be drunk without food. Where austere old world wines relied on food to soften them, the oak and ripeness did the softening here. If they had a fault, it was that they all marched to a stentorian beat; at every wine's core there were thumping bass notes.

Or so I thought. Like many who discovered Australian wines through strapping Shiraz and lush Chardonnay, I assumed they were all triumphs of substance over style, fine for enthusiastic quaffing but lacking the subtlety for measured appreciation. Many wines conformed to this stereotype, but to see the vinous produce of a continent-sized nation in such a one-dimensional way was to miss a wealth of treasure. The Aussies painted from a broader palette, more varied, more subtle, and more exciting, as I discovered in the Hunter Valley.

The Hunter is the Australian wine region that should not be: it doesn't rain when needed, in the growing season, and it does when not, at harvesttime. It is the historical cradle of Australian wine but only flourished because of its proximity to Sydney. Yet from these shaky origins emerged Hunter Valley Semillon, a wine with a cut-crystal flavor, announced by a lightning flash of acidity and not a whisper of oak. In youth, that is. For, carried within that monochrome flavor is a potential to develop and improve that few Chardonnay flavor bombs possess. It deepens with age, gaining an unforeseen toasty element—as I discovered on my visit to the Mount Pleasant winery where a tasting of the "Elizabeth" Semillon banished any set-in-stone notions about Australian wine.

Great Grape: Semillon

It is hard not to feel a certain sympathy for Semillon, the slightly awkward member of the grape family that never gets the recognition it deserves. When the Great Grapes gather for family photographs, Semillon is never given a place in the front row. It can sing many tunes and perhaps that is its weakness; it is capable of excellence at either end of the flavor spectrum, from bone dry to abundantly sweet. Such versatility casts it as a Jack-of-all-trades.

Sometimes it has indignity heaped on it, being heaved anonymously into blends as a filler or to get rid of a surplus. When Semillon is treated as a noble variety, however, it can deliver wines of memorable intensity and great longevity. For dry wines this happens most notably in Australia's Hunter Valley, and for sweet wines Bordeaux's Sauternes leads the field. Sauternes is Semillon's historic home, and its name may derive from the local pronunciation of Saint Emilion, *semeljun*, where a tiny amount of the grape is found. In Sauternes it yields one of the longest lived wines, which also happens to be one of the most unfashionable. Fashion is fickle, however, and may yet swing in its favor. It is too early to write Semillon's obituary.

Then winemaker Phil Ryan took the routine winery tour at a gallop—one stainless steel tank looks much like another—but the pace slowed when we sat down with nine vintages and spent the next hour tasting and retasting, comparing and assessing. First released as "Hunter Riesling" in 1967 and accurately labeled since 1982, "Elizabeth" is frisky in youth, a little awkward, but then it settles and develops in a way that defies its lean frame, filling and rounding over six or eight years to emerge as a classic. There's also a big brother wine—Lovedale—from a vineyard first planted in 1946. It has the same bone structure as Elizabeth but a fuller figure. More taut and closed when young, it develops remarkable amplitude after a decade or so.

Since that epiphany I have bought cases of Hunter Semillon from a host of producers, always delighting in its great value. The chalky-dry-in-youth, toasty-in-age character is highly rated by wine lovers yet largely ignored by the drinking public. On one occasion, having spotted a genuine supermarket offer and not a laughable "half price" whizz, I counseled my readers: "Do whatever is necessary to secure some of this wine: break open the piggy bank, spend the confirmation money, sell the car." I don't think there was a stampede.

The style is not middle-of-the-road. It has something to say for itself and it speaks clearly and without a heavy hit of alcohol. It is lissome but never frail; if you don't like its accent you'd best look elsewhere. Hunter Semillon is an Australian classic the like of which is not found anywhere else on the planet, and if there is a top dog among many contenders it is the Tyrrell's Vat 1, first produced in 1963. The youthful straw-green color barely yields to the passing years, and even when fully mature there is still a spark in the flavor. It is flinty clean with abundant citrus wrapping a mineral core, taut and pure, and eventually richer with resounding length.

It was perplexing that I had to travel to Australia to discover these Semillon wonders but the Aussies themselves were partly to blame, having peddled a narrow message about their wines, majoring on big hitters. Given the success of these headline styles they let the others languish and—whisper it—were perhaps keen to keep them for themselves and not share them with the world.

It is impossible to speak of Australian wine without mention of her most celebrated—Grange, formerly Grange Hermitage. My introduction to it was faltering thanks to a corked bottle, but a second was summoned from the cellar by Peter Gago, now chief winemaker for Penfolds. It was the 1992, generally regarded as a lighter Grange and, while I noted its restraint and good length, my pen didn't fly across the page to heap praise on the wine.

That came years later at a dinner I organized in Dublin, with Gago as guest of honor. The star wine was Grange 1990, shipped directly for the occasion, and the first to be labeled as "Grange." As if to celebrate its trim new identity, the wine was magnificent. The bouquet surged from the glass, a merry-go-round of brooding fruits, with a whiff of intrigue signaling a wine of many dimensions. Pyrotechnics followed on the palate: abundant fruit, leavened with coffee, chocolate, and licorice, and a savory note, all woven together by fresh acid and firm tannin. A big, generous wine, yet nimble too. For half a century, Grange has been acknowledged as a great wine, one of baroque splendor, yet it almost didn't make it out of the starting blocks. That it did was due to the vision and persistence of one man: Max Schubert.

Of German descent, Schubert was inspired by a visit to Europe: "It was during my initial visit to the major wine-growing areas of Europe in 1950 that the idea of producing an Australian red wine capable of staying alive for a minimum of 20 years and comparable with those produced in Bordeaux, first entered my mind." He set to work on his return, but with little Cabernet Sauvignon available he switched to Shiraz and made his first vintage in 1951. Five more followed and, while Schubert was pleased with his wine, management became increasingly concerned at the money tied up in maturing stock. Samples were called for, tasted, and declared dreadful. Some of the scorn heaped on the wine prompts incredulity

Talking Wine

It is the elephant in the room of wine communication: many winemakers who are wizards in the cellar, and loquacious when weaving a path through barrels and vats, are struck dumb when asked to present their wines in a formal setting. Clipped, strictly factual pronouncements replace the fruity cellar argot, reducing the audience to glassy-eyed boredom.

Step forward Peter Gago, chief winemaker at Penfolds since 2002, and a man who could give tutorials in wine-speak should he ever hang up his pipette. Gago, a former mathematics and science teacher, is an engaging and articulate speaker and I have yet to meet the winemaker who can elucidate a wine's qualities more eloquently than he. Making full use of his experience in the classroom, he communicates in a lively but never showy fashion, equal parts informative and entertaining. Attendees at his presentations always come away having gained some insight and understanding, without being swamped by a torrent of fact-based information that could easily be gleaned from a brochure. Whether speaking to the general public or a professional audience Gago always delivers—and he is no mean winemaker either.

today: "A concoction of wild fruits and sundry berries with crushed ants predominating." And: "Schubert, I congratulate you. A very good, dry port, which no one in their right mind will buy—let alone drink." Years later he admitted that his belief in Grange was shaken by all this opprobrium yet he persevered, even when the order to cease production came from head office in 1957.

Schubert's character was written in his granite features. Formidable determination and cussed self-belief were his calling cards. He was damned if company management, in thrall to the bottom line, was going to scupper his baby. Yet he needed every scrap of determination as he soldiered on, making a few clandestine vintages, until a few people began to take note in the early 1960s. After that, success came gradually and then in a surge of worldwide approval. Not only had he created one of the world's great wines but also one of its great wine tales.

Schubert's struggles are now part of global wine lore, recounted wherever and whenever Grange is poured. The wine is celebrated, sought-after, and fawned over—an "icon" wine—though the quip about whether "icon" is one word or two grows ever more apposite, thanks to the recent release of some ludicrously priced special bottlings. The reputation threatens to swamp the wine, for it is treated like a trophy, as if simply possessing it is an achievement. When another auction record is set for a renowned old vintage, it is reported in breathless prose under clickbait headlines. This is wine as dancing bear, performing tricks for a paying audience. At variance with this is the utilitarian label, a dull affair, muted red on muted gray, bereft of heraldic flourishes. It bears no relation to the razzmatazz that surrounds the wine and is a breath of fresh air compared to some of the ghastly eye-catchers slapped onto bottles of dubious merit today.

With Australia now acknowledged as a top-drawer producer, no longer the brash upstart, it is hard to imagine a time when that was not the case. One hundred years ago, production was dominated by fortified wines and strapping reds that delivered a sucker punch of alcohol and little else. Save for a few nuggets of excellence, the wines roared out of the glass with one whack of flavor and were gone. Australia was pigeonholed by them, but then came the decades of transformation, the 1950s and '60s, and by the end of the century Australia could do no wrong. Wine scribes flocked to visit, bounding from one winery to another like shuttlecocks; there was no more exciting country to explore.

Arriving in Melbourne after a three-flight odyssey from Ireland was not good preparation for a wine tasting two hours after landing.

A mix-up with dates had wiped the intended twenty-four-hour jet-lag recovery, replaced by a shower, coffee, and then winery visit and a slew of wines, tasted to the soundtrack of proud winemaker detailing the minutiae of production. I acquitted myself passably and it was only at the third winery that mind and body parted company, leaving me unable to process the data flood. The conversation drifted away then snapped back, drifted and snapped. At one point I went to speak but couldn't summon the words, which floated in my mind, beyond capture. I hoisted the white flag.

Assimilation is everything for the wine writer, shipping the torrent of information that comes in unrelenting waves, winery after winery. We force-feed on facts, gorge on figures. Free of jet lag,

Brilliant Pinot

If anybody can be said to have elevated Australian Pinot Noir to world-class level, moving it away from a too-hefty flavor profile, it is Phillip Jones, former owner of the Bass Phillip estate in Gippsland, southeast of Melbourne. I once described Jones in complimentary terms as "a Burgundian trapped in an Australian body." Equal parts charming and cussed, with a sorcerer's touch when it came to Pinot Noir, Jones founded Bass Phillip in 1979 and for four decades went his own way, inspired by great Burgundians such as Henri Jayer and Charles Rousseau. His best wines had an ethereal beauty—bereft of weight, density, and concentration—and such descriptors were redundant when it came to capturing them in words.

Times move on and in April 2020 Jones retired and sold out to a group of investors fronted by Burgundian Jean-Marie Fourrier. It is too early to cast judgment, but Fourrier is no slouch on his home patch of Gevrey-Chambertin. His wines are marked by purity and elegance, each a valid expression of its vineyard origins. Bass Phillip is in safe hands.

but with a packed schedule, it is a challenge to take it all in. At times it is a challenge to remember the name of the person you are talking to. Arriving at a winery, there might be a polite inquiry about where you visited the day before. Checking the itinerary is the only way to be sure.

The aim is always to glean something of the place that cannot be found in the brochure or on the website. The smell, the sound. Otherwise the visit is pointless and could take place virtually. I seek out the inconsequential details that a story can be hung on, pinpricks of light to leaven the technical tedium. Impressions are scribbled, especially the easily forgotten asides that burnish a story. Many Australian winemakers are strangers to crafted corporate blandishments, so keeping an ear cocked for throwaway remarks

The Ghetto Blaster

I once opined: "Little has been written about the unsung role of the ghetto blaster in the Australian wine industry, yet there is hardly a winery you can visit that doesn't have one, and it is usually balanced precariously on top of an oak barrel. At full volume in a cavernous cask hall they make an impressive din and render attempts at conversation in their vicinity useless. The cellar hands—often taciturn men who look as if they never leave their place of work—tap and sway to the rhythm as they go about their various tasks."

These sound systems are industrial grade, strong framed to withstand a fall from their perch. Frequently they belt out music, some say noise, to an audience of none, to what effect? Perhaps the thumping sound influences the wine, in much the same way as playing Mozart to children in the womb is supposed to turn them into geniuses before they even draw breath. If so, could this explain the booming Australian wine styles of the latter years of the last century?

brings rich reward. There's no anodyne magnolia-speak as they chat "off brochure," their observations laced with ribald asides. They favor the verbal straight right, especially when out of earshot of the chaperone from the marketing department.

Today we have a weapon that greatly assists in coping with the information tsunami: the camera phone. It is a black hole into which any number of pixels can be shoveled, so photos of no merit other than as aide-mémoires are snapped off, rapidfire. I photo bomb every winery I visit, picturing anything and everything. Short videos are the next step, with verbal observations that, if caustic, need to be out of earshot of the host. That is where the notebook is still handy, especially if you are with a group. You can scribble furiously as the winemaker speaks though you may be writing about how bored you are, for not all of them favor their visitor with choice quips and verbal fireworks.

They are not all motormouths. The late Dr. Bailey Carrodus, who founded his Yarra Yering winery in 1969, was one of the most reserved winemakers in Australia. As taciturn as the stereotype was brash, he let his wines do the talking and they spoke in their master's idiom, not the bells-and-whistles dialect that hollered for attention. Restraint was their signature, and flavors were deep, not expansive, built on solid foundations without flashy superstructure. Ditto the labels, black print on white paper announcing—in studiedly downbeat fashion—"Dry Red No 1" or "Dry Red No 2," the first a Bordeaux blend, the second a Rhône.

None of this prevented them from achieving cult status, and a "Stock Sold Out" sign hung regularly at the winery entrance. When I visited, the absence of chatter as the wines were tasted was disconcerting at first, so inured was I to talk-assisted tastings. At Yarra Yering you made up your own mind about a wine, without interference, and your opinions might be met with a noncommittal "Mmm" or a quizzically raised eyebrow. Or silence.

The odd clipped pronouncement broke that silence, though it might have nothing to do with wine, perhaps explaining the presence of a crate-full of marbles—used to "top up" barrels after samples had been drawn for tasting, making good the loss without needing a supply of wine for the purpose. The winery was boutique, but only in the sense of small, for it was unburdened by the flashy baggage associated with that label. Carrodus was a one-off who led the renaissance of the Yarra Valley, following years of devastation wrought by phylloxera and a change in the national taste in favor of fortified wines. Since his death in 2008 and the subsequent sale of Yarra Yering, I cannot imagine that a visit is as interesting, if unnerving, as previously.

The scale and nature of the Yarra Valley make it an easy place to explore and a hard place to leave. Yet no visit to Australia is complete without a stop in her wine industry's engine room—the Barossa Valley. There's a Teutonic feel there, a legacy of its settling by German-speaking immigrants from Silesia in the mid-nineteenth century. The buildings are solid, the food is hearty, and they are

Bring Your Own

The Yarra Valley is Melbourne's back garden, supplying her sophisticated dining scene with wines to match a cuisine that makes the city a gourmet's paradise. Many of the restaurants operate a bring-your-own policy, a circumstance I took memorable advantage of after a week visiting wineries across Victoria. I normally always refused the offer of a bottle at the end of a visit but, with my wife working as guest concertmaster of the Melbourne Symphony Orchestra, it was easy to whistle up a dozen musician-diners for the Friday evening. I staggered into the Saigon Rose restaurant, carrying a mixed case of wines. "Which should I open?" asked the waitress. "Open them all," I replied. The bottles were laid in a line down the center of the table. Thereafter, whenever my wife was invited back, another inquiry quickly followed: "Will your husband be with you?"

matched by the wine for which Barossa is famous: no-holds-barred Shiraz. If Mosel Riesling is a Bach sonata, Barossa Shiraz is a Mahler symphony, grand in every dimension.

And yet there was a time in living memory when the ancient vines that yielded this heady brew were barely valued. Some were over a century old and a few were probably planted by the settlers—and tended by generations of their descendants. These farmers relied on selling their grapes to the large wine companies whose reach was national and who could source cheap fruit from across the country for their price-driven blends. They cared little that the vines had never been troubled by phylloxera, that the wizened veterans were some of the oldest vines in the world. When they chose not to pay a fair price, indicating they were content to let the grapes wither, ruin beckoned. Where phylloxera failed it looked as if the vines would succumb to the hand of corporate man.

Which was to reckon without the resolve of one man, a human bulwark who stood for the growers and made it his mission to save their vines and their livelihoods. Peter Lehmann, a true character in a character-heavy industry, was a gruff, chain-smoking, and opinionated advocate for all things Barossa. It was easy to imagine Shiraz flowing in his veins. He lived beside the eponymous winery he founded to buy the unwanted grapes and during harvest took station at the weighbridge to shoot the breeze with the growers arriving with their grapes. Visitors to the house were treated to his uncensored wisdoms in the smoke-filled kitchen, where he held court with his wife Margaret. Though marked on no map, that kitchen was a Barossa crossroads, where a stream of callers dropped by for a chat or to bounce a nugget of news off the "Baron of the Barossa," often to have it shot back with a stinging riposte.

Lehmann loved full-throttle, whopping Shiraz, but since the turn of the century that hefty stereotype has yielded to a more nuanced profile. It once "deafened" the palate in the same way as loud music batters the eardrums; the music was good but the volume made it

The China Syndrome

Australian winemakers and marketers were always noted as sure-footed operators. Clumsy attempts to get consumers to buy their wines were for less savvy, more hidebound nations. And yet, in recent years, as Chinese demand for Australian wine exploded, they lumped all their eggs in one basket, falling over themselves to supply that vast market. Dazzled by easy success, they forgot the markets that made their name on the world stage, that gave them the calling card to knock on China's door.

Tiny Ireland was once Australia's sixth-biggest export market, after behemoths such as the United States and the UK. Yet when market share slipped, little was done to address the situation; there were bigger fish in the sea. It was a huge mistake to focus so intently on China, an uncharacteristic howler only realized when China slapped massive import tariffs on Australian wines in 2020. Timbers were shivered and hopefully the need to nurture all markets, even the ones no longer performing rampantly, will have been noted. Hopefully.

difficult to tell. Today, the best stuff delivers an avuncular embrace rather than the bear hug of yore.

However appealing Barossa is, I've always had a fondness for the contiguous Eden Valley, to the east and at a higher elevation. With a little altitude, the air is fresher and the wines too, including Henschke's Hill of Grace Shiraz, a serious challenger to Penfolds Grange. The crucial difference between the two is that the Henschke comes from a single vineyard while the Penfolds is a South Australian blend. Where "Grange" is the wine name, "Hill of Grace" is both wine and vineyard, and is the leading example of how place is becoming more prized in Australia, where varietal labeling is the norm. A wine's origin, where the vines' roots delved for sustenance, is increasingly likely to be indicated, especially on premium bottlings.

Hill of Grace is a remarkable vineyard, populated by antique vines whose trunks look like twisted elephants' legs. Ragged with age, they sag under their own weight, yet the wine coaxed from them is anything but ragged and now changes hands for knee-weakening sums. Henschke celebrates origin; Grange celebrates winemaking. Both wines are superb and demonstrate there is more than one path to excellence. In general, I favor origin over winemaking, but with a glass of Grange I eat my words.

Hill of Grace is Eden's flagship, but the valley's best Rieslings are also excellent and are site-specific too. The Steingarten vineyard was planted in the 1960s and occupies a hilltop site, its green foliage contrasting with the dun of the surrounding hills dotted with the odd tree. The breeze clips the ankles, dispelling thoughts of lingering, and the wine echoes its austere origins with a high-voltage thrust of acidity running through the trim fruit.

A gentler Riesling hails from the Pewsey Vale Contours vineyard, named for the vine rows that trace their way around the hillside like contour lines. This produces one of Australia's finest Rieslings, capable of aging twenty-plus years without faltering. The 1999, sealed with screw cap and drunk in 2020, was still bright-colored and brisk-flavored, shot through with notes of lime and a whiff of honeyed age. Eden Rieslings vie with those of Clare Valley for the crown of Australia's best. I lean toward Eden, preferring the friendlier profile; the acidity in Eden tingles, thanks to the redeeming pull of floral notes mingling with the lime, where in Clare it can shriek. Some Clare Rieslings skate across the palate while Eden's glide.

Clare Valley was settled by Irish immigrants in the 1840s who named it after the eponymous county back home and, though only a short distance northwest of Barossa, it feels remote and isolated. This engenders a strong sense of community, seen some years ago when producers, tired of cork taint clobbering their racy wines, took the collective decision to switch to screw caps for their Rieslings. Only by taking the leap together could they be sure to get the message across about their belief in screw caps. What was

All That Glitters

The everyone's-a-winner spirit of school sports days, with medals dished out like confetti, is all too evident on the Australian wine "show circuit"—a long roster of trade shows where wines compete against each other for medals of one shade or another. I put little trust in them, for far too many bottles end up festooned with the sticky little discs and I always ask the question—who didn't enter their wines in the competition?

There is a bewildering number of prizes available, but I can't help feeling they help the producers to market their wines rather than assist consumers in choosing between them. And if it leads to the winemakers trying to second-guess the judges and craft a wine designed to please them, then it is warping the business. Having only one eye on the fermenter with the other on the prize is hardly a good way to work. Too many medalists are underwhelming, the glitter fading once a glass or more is drunk. They may dazzle in big tastings but then collapse under the pressure of sustained scrutiny when drunk with a meal, or when age wipes the gloss of youth away to reveal nothing underneath. Medals serve producers, not consumers. Beware medal madness.

revolutionary then now seems like common sense and it was also a practical move, as it generated a large enough order for the new type of bottle required.

Clare may feel remote, but the truly remote Australian wine region is Margaret River in Western Australia, a continent's breadth away. Wine has been made there for little more than half a century, yet the Cabernet Sauvignons are rated as some of Australia's finest. They possess a restraint not always associated with Australian reds, but wholly in keeping with the chic milieu of the region. Did one beget the other? It is tempting to think so, for the swagger that is charming elsewhere would be out of place here. A measured stride is more suited to Margaret River.

Cabernet is king here, producing structured rather than powerful wines, ripe but not so ripe that Cabernet's savor is lost. Freshness bestows enduring attraction. You take a sip, then another. Whether solo or blended with Merlot and other Bordeaux varieties, these are winning wines. Sauvignon Blanc and Semillon is another favored blend, but where the Cabernets bear comparison with Bordeaux—a Leeuwin Art Series once called to mind a St-Julien—the Sauvignon-Semillon blends don't have the intrigue of the finest Graves.

Margaret River lies about two hundred miles south of Perth, and few regions are so richly endowed with life's finer attractions. There's natural beauty, artisanal food producers, bijou shops, and delicious wines; there's even superb surfing for gourmets keen to stay trim. And therein lies a snag for winery owners. Proximity to the Indian Ocean creates a unique problem for, when "surf's up," workers vanish. For most of the year it's not a problem, but at harvesttime when every hand is needed it is frustrating for a winemaker to discover the majority of his pickers have downed tools and headed for the ocean.

Can one wine, one single wine, transform a nascent wine industry into a player on the world stage? In the case of New Zealand, yes, and that wine was Cloudy Bay Sauvignon Blanc. It arrived out of nowhere in the 1980s and it is impossible to appreciate the stir it caused unless you experienced it firsthand. Each new vintage prompted a rush to secure a few bottles, and only when funds were exhausted would friends be tipped off. If they were quick—"drop everything" quick—there might be a few left. The irrational hunt elevated the wine to cult status. Nothing like it has been seen since.

It was vivacious and bursting with life. It danced a sweet 'n' sour quickstep as the flavor oscillated between complementary poles, from peach to gooseberry, a back-and-forth dazzle with taste buds happily caught in the middle. Cloudy Bay's shooting star raced across wine's firmament, snaring a legion of wine drinkers who couldn't get enough of it. For palates jaded by the usual flavor roster, it was a never-seen-before sunrise. This was new, not a shifting of familiar components to form a newish pattern. There had been a straw in the wind—Montana's Marlborough Sauvignon Blanc—but it was Cloudy Bay that introduced New Zealand to the wine world.

The heat of those days has cooled. Cloudy Bay is now one of a pack, its original winemaker, Kevin Judd, gone to produce a rival called Greywacke, but it opened the door for a legion of other wines. Perplexingly, it sprang from a nation noted for a disapproving relationship with alcohol. One hundred years ago New Zealand toyed with Prohibition, only narrowly missing that disaster, while still imposing myriad restrictions on consumption. Ally that to a strait-laced society, a fading glimpse of which I caught—staid tearoom straight from "Miss Marple," Wolseley and Austin motorcars—and it's a wonder anything so exciting emerged from it.

Prior to that, the winemaking landscape was dominated by dreary Müller-Thurgau, the grape-crossing progeny of a Doctor Müller from Thurgau, from which few wines of note have ever been made. Paradoxically, this was an advantage, for, with little winemaking tradition, the pioneers had a blank canvas on which to work.

Today, that canvas is adorned by a quartet of excellence: Riesling, Chardonnay, Pinot Noir, and Syrah. Australian Rieslings and Chardonnays used to hog the Antipodean limelight, but their Kiwi cousins are now equally fine, especially the Chardonnays. Pinot Noir, never Australia's strongest suit, is now New Zealand's red standard-bearers, sure-footed after shaky beginnings when, in this rugby-obsessed land, Pinot was made to play in the forwards when it yearned to run with the backs. I was reminded of the old days by a Wild Earth Pinot Noir 2006, drunk in 2020: a blocky wine untamed by the years, with none of the beguiling spark that is Pinot's signature.

Kiwi Pinot now sparkles and thrills, its only sin being the way it obscures from view the wonder that is New Zealand Syrah. This is

Great Grape: Syrah

Say: "Syrah is the greatest red grape," then duck as the Pinot Noir and Cabernet Sauvignon fans howl heresy down. Yet the evidence is compelling for, while Pinot makes the finest unblended reds, and Cabernet majors in the greatest blends, Syrah challenges each in their favored roles. Standalone Cabernet or blended Pinot don't have the same appeal. Syrah combines the grace of Pinot with the structure of Cabernet.

Its home is the northern Rhône Valley, where its fame rests on the twin pillars of Hermitage and Côte-Rôtie, supported by their attendant appellations, Saint-Joseph, Crozes-Hermitage, and Cornas. Here, Syrah sketches a chiseled profile; iron filings lurk in its depths while potent dark fruit and spice swirl above. In Australia's Barossa Valley Syrah beefs up and changes its name to Shiraz. The vines there are heavy-trunked veterans whose bark looks like fissured stone, and the wines are often sumptuous and luscious. Across the Tasman Sea in New Zealand, it's back to "Syrah," and a style that cleaves to the Rhône template. No matter what it is called this great all-rounder, aided by global warming and its own class, has time and trend on its side. In years to come historians will label this the Syrah century.

a wine of reserved joy, definitely Syrah, not Shiraz. Precise, elegant, and refined, with a firm handshake and a properly knotted tie. Much as I love the Pinots, Kiwi Syrah sings purer, with spicy savor rather than fruity tingle. It's slim, not bounteous, dry but not swingeing, lip-smacking and deserving of more recognition. It boasts a "coolness," a lack of aggression, that invites repeated sipping. It took only one example to convert me, and that was Man O' War Dreadnought Syrah 2008. The dry savor of dark chocolate mingled with black olive and pepper and bacon, with the fruit on the margins content to play a subsidiary role.

Yet, no matter how bright the quartet shines, Sauvignon Blanc still rules the roost. It grew into the cash cow that kept giving and it still is, now yoked to a host of bland brands, technically correct wines with cloned, watery calling cards. Boredom is threatened by this universe of sameness, these shaky imitations of a once-true original.

Flying home from my first visit to Australia and New Zealand, though not on *Wunala Dreaming*, I noted: "Aged Semillon is certainly the find of the trip, but I feel it will remain a niche wine. It doesn't provide easy, inoffensive, undemanding drinking; you have to pay attention." This remains as true now as it was then.

7

WHEN MONSTERS ROAMED

The wine world has seen its share of monsters, warped wines tugged out of kilter by misguided notions of quality. For a period their heft elbowed less-hearty wines aside as they bounded from the glass, suffused with richness and bereft of subtlety. They were children of the 1980s, when extravagance at the expense of elegance chimed with the "greed is good" mindset of the 1987 film *Wall Street*. It was a potent symbol of the era, greed *was* good, and big 'n' brash wines were needed to fuel it. Both movie and wines reflected their times, but fashion changed and today the blunderbuss wines are relics of a receding age; some cling on, appealing only to palates so blasted by torrid flavors that nothing less makes an impression.

The 1980s validated the excesses of the 1970s, outlandish fashions took center stage, and wine was a willing victim, bulking to a degree that rivaled the gargantuan shoulder pads and the billowing hairstyles of that decade. The list of victims reads like a roll call of the world's greatest wines: Bordeaux, Burgundy, Rhône, Napa, Tuscany, Priorat, Barossa . . . on and dismally on. The headlong rush for wines with bolted-on flavors and textures knew no limits. It was like disfiguring an elegant building with a clumsy extension, replacing harmony with discord. The result was a hectic amalgam of flavors that cudgeled the taste buds: barging fruit, thumping tannins, and brutish concentration, all laced by searing alcohol. Anything frail got trampled in the rush, outrageous flavors chimed with the zeitgeist; as the twentieth century closed only big hitters were welcome.

Most commentators ascribe their rise to the influence, baleful or otherwise depending on your point of view, of the American critic Robert Parker. That is a blinkered view, considering wine in isolation from wider society where the movers and shakers hollered for biggest and loudest. Having cut a billion-dollar deal, you celebrated with wines that grabbed the attention, seizing the palate in a wrestler's hold, not ones that couldn't get noticed for the raucous din and the cigar smoke. Wines were concocted to

Triple B

"Big but Balanced" is the argument trundled out with tiresome monotony in defense of monster wines that deliver a concussive whack of alcohol. If the wine is balanced, if all components are present in equally hefty measure, then their outlandish concentration will balance the high alcohol. If everything else is punching big, then the alcohol hit is excusable. In that scenario it won't be noticed until it delivers a rabbit punch hangover. A balanced monster sneaks in like a Trojan horse, while the unbalanced one is easily spotted and can be avoided. Excusing all excess with the glib "Big but Balanced" response should be the preserve of those tasked with marketing and selling the wines, not those commenting on them; they should call them out for what they are. Balance does not absolve all sins.

match the preening egos as they chest-bumped and marveled at their genius. Wines that required quiet consideration were for wimps.

A blinkered view also ignores the landscape in the pre-monster age when too many long-established classic wines were only intermittently classic, occasional Thoroughbreds, frequent donkeys. Too often they had a bag-of-bones character at odds with their gilded reputations. They were ripe for changing. Thus Parker was pushing an open door when he celebrated lavish flavors over weedy ones. By the turn of the century wines resembled the graceless 4x4 automotive barges that clog roads and clatter around parking lots, bumping and scraping to left and right.

Think of wine as an orchestra, a harmonious combination of components. Paramount is the overall effect, achieved by judicious use of each component. Stretching the analogy, it was as if the violins were ignored in favor of the double basses; the flutes in favor of the trombones and the percussionists were encouraged to belt away as they pleased. Boom replaced delicacy. Wines delivered an avalanche of flavor, not a flowing river.

Winemakers became caricaturists, cherry-picking flavors and exaggerating them wildly, as the political cartoonist distorts the prominent nose or bushy eyebrows of his subject. Some grapes were better able to handle the rough treatment than others: Cabernet Sauvignon and Syrah/Shiraz could survive the bully boy winemaking and still come up smiling, if bruised. Ramping up Cabernet's signature blackcurrant fruit and stern tannin presented few problems; ditto with Syrah's spice and savor, which was volume control more than anything else. Once the grapes were ripe or, better still, over-ripe, you got away with it. Pinot Noir was trickier; delicate-but-intense is harder to bloat, so the worst examples ended up tasting vaguely of Pinot with heftier notes teetering on top. It was Pinot cast as Cabernet. And any Sauvignon Blanc that received the treatment wobbled along for a bit before collapsing in a heap of opulence far removed from the grape's signature zing.

The monsters were essays in warped winemaking, superficially impressive but under the gloss the wines were awkward and heavy-hoofed, like athletes that went to the gym via the backstreet pharmacy. In all this, winemakers were aided by advances in winery equipment that encouraged the march to excess. Blinded by the possibilities, they produced vinous monuments that concussed the taste buds. There was too much winemaking and not enough respect for the raw material coming from the vineyard. The wines were branded with the maker's mark; nothing was left to chance, it was like a chef adding too many ingredients because he doesn't have the confidence to hold back. What you leave off the plate is the tough decision. Ditto wine, it is easier to keep interfering than to stand back and observe. That takes confidence.

Super Sorting

It is heretical to suggest, but perhaps the contemporary craze for super sorting of grapes, where bunches are scrutinized on an almost grape-by-grape basis to weed out anything deemed substandard, is taking things too far. In pursuit of flawless wines, only faultless fruit gets past the winery door. But when flawlessness is championed ahead of character the results are admirable rather than engaging.

They encourage arms-length appreciation; it is easy to be objective about them, coining measured comments and cerebral opinions. Unlike the wines that get underneath your skin, welcome invaders of your spirit that dare you to remain objective when your every desire is to tumble into subjective revelry. Their siren call is irresistible and it is often because of a quirk, a flavor inflection that some would deem a fault. Is super sorting simply a way of giving wines a too-perfect polish? Perhaps we have seen enough of it.

The more-is-better philosophy was underpinned by a vineyard practice that gave the mantra du jour, solemnly enunciated in tones that belied its fad-of-the-moment status: "hang time." This involved leaving the fruit on the vine way past the point of proper ripeness, with sugar increasing and acidity plummeting, until the grapes turned into sugar sachets, to be fermented into syrupy wines. The practice rid the wines of any "green" flavors, but it was a baby-with-the-bathwater exercise. How long you left your grapes on the vine after they should have been picked was spoken almost as a boast, like saying how much you bench pressed in the gym. It was a false god that delivered counterfeit class. When this craze was allied to extra time in new oak barrels, the wines were burdened with a load they could not carry.

It would be impossible to overstate the importance of the oak barrel to the wine industry. What started as a storage vessel in Roman times, robust enough for transportation, was then discovered to have a significant influence on the wine inside. At its simplest it adds flavor to the wine, like seasoning in a dish, and it is this quality, more than other complex influences, that sees it used and overused in so many wine regions. The trouble starts when it becomes an ingredient rather than seasoning.

For centuries the barrel has hardly changed and its method of production, while now utilizing some machinery, still relies on skilled artisans whose strength and dexterity transform jumbles of staves into smooth barrels, each held together by ingenuity and half-a-dozen metal hoops. Visiting a *tonnellerie* or cooperage is not for the fainthearted. Outdoors, stacks of rough-cut staves stand weathering, leaching tannin onto the ground below, creating a drab, gray-brown vista. All the action is indoors, where the eardrums are assailed by staccato hammering, as hoops are driven onto staves to yield a rough barrel shape. Earplugs are de rigueur, conversation is barked. Some two dozen staves are used for each barrel, and the pace is relentless. The cooper doesn't pause for thought or scratch his head; he's done this before.

When half-made the barrel is ungainly, with staves held tight at one end and splayed awkwardly at the other. Fire and water and a steel hawser are used to bend them to shape, allowing more hoops to be added. Then come the top and bottom and, apart from the drilling of bungholes and some sprucing up, the barrel is complete. Machine-gun hammering, groaning timber, dancing flame, smoke, and steam combine to set a chamber-of-horrors scene that transforms a clutch of ugly gray staves into an elegant barrel, smooth-cheeked and shrink-wrapped for transportation.

The choice of timber is paramount and French oak still holds the high ground, revered for its subtle influence compared to the more strident American. Swathes of France are covered in oak forests, many planted originally for warships and, with a gap between planting and harvesting of perhaps 150 years, they are managed by the state. The trees are trained to grow straight and to a height of about 65 feet. Each tree will yield no more than two barrels whose useful life is about three to five years before they descend down the food chain, eventually ending up as garden furniture and then fuel for the barbecue.

When sensitively used, oak barrels add immeasurably to the quality of a wine, framing the picture, revealing and giving emphasis to intrinsic qualities. But when thrown at a wine in a more-is-better fashion they trample it, obliterating subtlety. Thus used, they are an expensive indulgence for which the customer pays doubly—through wallet and tattered taste buds that struggle to discern some flavor through the oak fog. In the case of a feeble wine, more shadow than substance, a faux elevation of quality can be achieved by "garnishing" it with oak chips. This involves suspending a "tea bag" filled with chips in the wine, to paper over cracks and fill in for what is absent. No matter what the sales guff might claim, an ordinary wine cannot be transformed into something celestial in this fashion.

Spain has the daddy of all affections for oak barrels, in many cases American, which lavish the wines with vanilla and spice. Winery visits in Rioja or Ribera del Duero always include a proud mention of how many barrels they house. Yet that is like describing a house by saying how many bricks it contains; it is irrelevant. Too often I have visited wineries there and been impressed with the "standard" wine before finding the "superior" wine less enjoyable, its profile warped by oak. It's like applying a heap of makeup when a touch is fine. Oak is foundation for wine, not superstructure.

Vega Sicilia

Long before Ribera del Duero's rise to fame, Vega Sicilia was its sole winery, turning out its eponymous wine since the 1860s. Despite recent competition it remains unchallenged as Spain's greatest, though a recent visit to the winery remains unchallenged in my memory for the frostiness of the welcome, the barked orders from the security guard forming a grim contrast to the quality of the wine. By contrast, our guide was all sweetness and light but that failed to erase the initial, heavy-handed impression.

The wine is far from heavy-handed despite many years oak-aging, a practice that would destroy most wines. In this respect it is the exception that proves the rule. Robust yet refined, with fathomless depth of flavor, it is a peerless beauty, a unique statement of winemaking excellence. The best vintages need two decades aging before their tannic cloak softens. The 1990 "Unico" drunk in 2016 alongside a Château Haut-Brion 1990 remains a favorite: "meaty and vigorous with a rustic edge, intense and long, lovely sharp tingle of spicy fruit, incredible depth of flavor, heartier than the Haut-Brion." There is also a "Reserva Especial," blended across three vintages, making it a contender for the title of world's greatest nonvintage wine.

Peak oak was reached when some rabid imaginations struck on the idea of "200 percent new oak," a practice that involved aging the wine for perhaps nine months in new barrels and then transferring it into a fresh set of new barrels, resulting in a wine-and-oak consommé. It was akin to dressing in one new suit on top of another and expecting people to be impressed. Mystifyingly, some were. Such was the craze for lavishing wines with new oak that the barrel came to be seen as the villain, but it's a passive instrument. It did not assault the wines unaided; it was set upon them by winemakers surfing the tide of fashion.

More oak plus more hang time plus more of everything else was blunt force trauma for wines and gave birth to a new, international style that plodded on the global stage. The style was easily identified but not the origin. It was big and suave with rich flavors, probably made from Cabernet Sauvignon. But made where? These were "anywhere" wines, superbly crafted citizens of the world, rooted nowhere, identifiable as "serious" wines by lavish flavors that tasted mainly of expense. Manicured in the cellar, they were deep and resonant, and warm on the finish: undoubtedly impressive, dubiously satisfying.

As a rule of thumb, the more prestigious the wine the more likely it was to get a supersizing makeover. Bordeaux fell over itself in the craze for new oak, forgetting that some of the wonders from decades past were made with fewer new barrels, because they could not be afforded. Ranks of pristine barrels stretching into the distance became the norm in the *chais*, or barrel halls. Bordeaux is the standard-bearer for fine red wine worldwide so when it embraced excess others followed, beefing up to chase critics' scores. Yet claret should be refreshing, and freshness is not a bedfellow of concentration. Without freshness it is indistinguishable from the hordes of "me too" Cabernet-based wines that hang on its coattails. Proper claret meets the taste buds with a handshake, not a punch. It shouldn't dazzle.

Bordeaux's big-is-better standard-bearer was Château Pavie. Once a gracious wine, it had a personality change foisted on it and emerged disfigured and barely recognizable. The 2003 drunk at fifteen years old sticks in the memory, as it did in the throat. It was spicy and brusque, the flavor unraveling when it should have converged into elegance. It lacked the sweet savor of aged claret, a flavor focus around which others array. It was far removed from the 1982, drunk at about the same age; gentle persistence was its calling card, not hollering insistence. All they had in common was the name, like a father and son where the father was the epitome of civil restraint and the son was a coarse blowhard.

Meanwhile in Burgundy, "pre-fermentation cold soak" was championed as a fix-all remedy to get more stuffing into feeble wines. It was an understandable development, as the region recovered from decades of overtreatment in the vineyards, resulting in wines of scrawny charm. The practice of holding the must at a low temperature to delay fermentation wrung the grapes for color, tannin, and extract, and answered the call for substance at all costs, pulling harsh flavors and coarse textures from skins and pips.

Farther south, one of the worst offenders was Châteauneuf-du-Pape, the southern Rhône's standard-bearer. Special bottlings were its undoing, best-of-best selections with concentration prized above all. Proper Châteauneuf needs no embellishment; it's already a baroque wine, from plangent name to embossed bottle, sweet and spicy, leathery with age. The best wines are laced with a *sauvage*, herbaceous streak that prevents plod and sustains interest. Poise is maintained even when opulence abounds. Plastering Châteauneuf with lumpen flavors insults its natural substance; it doesn't need amplifying. In time, the best-of-best aberrations will act as a warning to future generations not to be wooed by the promise of easy success through shouting louder. Cuddly warmth counterpoised by a peppery ping is the ideal.

Worldwide, it wasn't only red wines that suffered these indignities. For a period, South African Sauvignon Blanc had a

Great Grape: Grenache

Also known as "Garnacha," to give it its more correct Spanish name. Despite being a sun worshipper Grenache knows all about shade, for it lives in the shadow of Syrah in France and Tempranillo in Spain. These garner the plaudits while Grenache toils in the background, providing ballast in blends and seldom getting the chance to shine in its own right. When consigned to such workhorse duty it contributes two essentials, sweetness and alcohol, sometimes in large measure, when they intrude on the wine and swamp other flavors.

As a champion blender Grenache provides the foundation for oceans of Côtes du Rhône. When finally given a chance to shine—as in Châteauneuf-du-Pape—Grenache can come across as a bit of a lush, heavy on the costume jewelry. Yet it is those qualities that give Châteauneuf its unmistakable character. Grenache shines in this papal incarnation and shows itself well capable of playing in the top league on occasion—as in the magnificent Château Rayas, which is made from 100 percent Grenache.

torrid time, as the weight was piled on to its delicate frame. There was some appeal on the nose but the palate was murky; it was akin to thickening a delicate consommé. After Sauvignon had been to boot camp it emerged with sloppy, blowsy flavors, like artificially sweetened coffee. South African reds were better able to handle the rough treatment, coming to resemble the rampaging rugby forwards the country is famed for. It was paradoxical that fleet-footed Sauvignon suffered the worst treatment. The good news is that it has bounded back, emerging as a world-class example of a grape that is often one-dimensional. The steroids have been forsworn and the style, particularly from Constantia, charts a fruity-mineral path between tear-jerking intensity and over-opulence.

Back in Europe, the malaise knew no bounds. Forty years ago in Spain, Ribera del Duero bounded onto the wine map and stole Rioja's thunder, going on to boom loud and long. The wines were lush and dense, Rioja with knobs on, and for a period they ruled the roost—arriviste to Rioja's ancient reserve. The elder statesman of Spanish wine was caught napping as Ribera bellowed its charms, lacking only subtlety and reserve, elegance and finesse. Caught in the headlights Rioja lost direction, while the new kid on the block wowed critics and consumers. But Ribera del Duero tripped itself up. Gorged on success, it troweled on weight, ratcheting up the density to the point where recently it was easy for Rioja to slip back into the limelight, like a nimble footballer sidestepping a trundling defender.

Great Grape: Tempranillo

Is Tempranillo the Switzerland of grapes, respected rather than loved? It's a plausible assertion, yet it misses the point, for Tempranillo is worthy of its "great" status, not because of a strident personality, but because of its opposite. Tempranillo—especially in its Rioja manifestation—is always pleasant company; there aren't any flavor pyrotechnics.

As a result, it never gets the attention it deserves, a situation compounded by Tempranillo's name changes as it travels the Iberian Peninsula: Ull de Llebre in Catalonia, Cencibel in Don Quixote's La Mancha, Tinto Fino in Ribera del Duero, Tinta del País in Castile. Across the border in Portugal it is Tinta Roriz in the Douro Valley, while farther south it answers to Aragonez. Rioja is Tempranillo's heartland, where traditionally it delivered a heady brew of savory fruit lavished with the vanilla smack of American oak. Good examples age remarkably well; top drops can handle half a century in the cellar and still come up smiling. In this respect it can often outpace more storied wines such as Bordeaux, a fact not often acknowledged.

Ribera's behemoths stand like Stonehenge wines, hopefully not for as long.

The situation in Italy was different. There, the monster malaise was fed by the success of some adventurous though not excessive wines, before taking flight to yield a plethora of waddling wonders that gained inexplicable acclaim. The unwitting progenitors were the likes of Sassicaia and Tignanello from Tuscany, Italy's best-known region and home to her best-known wine, Chianti. As such they represented a refreshing change from the drab Chianti that was commonplace in the 1970s. It was easily shouldered aside, for it was a weedy libation, for sloughing back in tandem with pasta or pizza to ameliorate its tart rasp.

Time spawned a legion of bandwagon jumpers and they, along with the originals, came to be known as the Super Tuscans, an apt designation that sounded like a TV crime drama. Initially there were some wonderful wines but then, what could have been a glorious reinvention, was hijacked and set on a clumsy course in search of greatness. The newcomers feasted on fame, growing ever more grotesque; they lost the run of themselves and soon became indistinguishable from countless other heavyweights being churned out across the globe. It would be wrong to tar them all with the same brush, for Sassicaia and Tignanello have retained some of the grace and elegance, resting on a firm structure, that set them apart about half a century ago. Theirs was a laudable initiative with delicious results but their successors spun out of control, blinded by success.

Perplexingly, farther north in Piedmont, Italy's greatest wine, Barolo, soldiered on, not always immune to excess but undazzled by the limelight that glared on Tuscany. Granted, Barolo's sibling, Barbaresco, saw some bulking up courtesy of Angelo Gaja's shout-from-the-rooftops approach, in both winemaking and marketing. The worst of that is behind us, and Piedmont's best wines are now accorded their due, thanks to consumers seeking out more refinement and less lushness in their wines. Barolo fits that bill. It's never "easy." Austere in youth, it is only with age that singing

Great Grape: Sangiovese

Tuscany—font of dewy-eyed reveries among all who have been there and all who haven't. Tuscany—home to Chianti, with Florence to the north, Siena to the south and, in between, the sylvan slopes, the classic villas, the hilltop towns . . . and wine, oodles of wine, much of it made from Sangiovese, an easier grape to appreciate than to love. Too often there is a brusqueness to the flavor that jars on the palate and sets a limit beyond which few wines go, despite careful cultivation and dextrous handling. Where, for instance, Cabernet Sauvignon regularly reaches magnificence, Sangiovese spends a lot of its life not quite getting there.

The "Super Tuscan" movement blended Sangiovese with "foreign" grapes such as Cabernet to produce wines that garnered huge acclaim, though eventually they morphed into lavish brews with little Tuscan identity. Their arrival did, however, have the knock-on effect of boosting Sangiovese's fortunes and hence the quality of Chianti. Some say it challenges Nebbiolo as Italy's greatest grape. I demur.

sweetness emerges out of the cloak of tannin. It deserves time and rewards time.

Big wines needed big bottles, bottles that spoke of their contents' character before a cork was pulled. And how they spoke. Wine bottles ballooned, keeping pace with the exploding waistlines of the wines they contained. Many became grossly proportioned. They were— and still are—brutish beasts. Like today's trunk-necked sportsmen with Popeye biceps, they make their predecessors look scrawny and underfed.

The weight of these bottles is obscene, and while others bemoan their impact on the planet my worries are more selfish.

What if I drop one on my toe? Steel toe–capped boots should be issued to sommeliers required to handle them—to save on personal injury claims. They look like liter bottles yet only contain 750 ml and need two hands for safe pouring. It's possible to pick up an empty one and attempt to pour some wine, the weight suggesting the bottle contains some. They are also a nuisance to store. Cellar bins constructed to house traditional bottles struggle to hold their gross girth. Those with tapering rather than cylindrical sides may look elegant when upright but form a jumbled heap when they are stacked on their sides.

Not long ago a trio of shapes ruled the roost: the cylindrical Bordeaux bottle, the curvy Burgundy bottle, and the swan-necked German bottle. A modest flight of imagination suggests each matched its contents: the first reserved, the second sensual, and the third delicate. Bordeaux now wears shoulder pads, Burgundy has gone dumpy, and Germany has grown alarmingly tall. Some are freak show caricatures of the originals, others have retained elegance, utilizing emphasis not exaggeration to render a contemporary re-working of a classic, expanding graciously without throwing the traditional shape off kilter. Those sacrificed at the altar of grossness added weight, weight, and more weight, resulting in brutes of graceless mass, body armor for wine.

Someday, I hope to hold a "Heavy Bottle Tasting" which would be "palate blind," meaning that, unlike normal blind tastings where the bottles are carefully wrapped to disguise their identity, these bottles would be sighted but unopened. The tasters could do everything except taste the wine. Feel the weight of the bottle, examine the label, and note any lavish embellishment, search the back label for purple prose, check the alcoholic content, write the tasting note. Only then would the bottles be opened for actual tasting to check the accuracy of the notes. It would be discovered, *quelle surprise*, that it is easy to paint a broad brushstroke tasting note. Then comes the good bit: having tasted the wines and refined the notes, comparison is made between all the wines to see how similar they are. They may be from all over the globe, red, white, whatever, but they will be found to share many general characteristics, remarkable similarities of style if not actual flavor: the heavy bottle style.

Just as the Botoxed face wears a dazzled expression, so too do over-worked wines have a "dazzled" flavor, locked in place, impressive rather than attractive and not receptive to interaction with the drinker. They are monoliths—all subtlety is expunged in the winery. You appreciate them rather than interact with them; all that is required of the taster is dumbstruck admiration.

Mercifully, the worst days are behind us, particularly the use of new oak. After recent excesses, oak is returning to its proper place as an aid to winemaking, resulting in wines that are fresher, more appealing, and less ponderous on the palate. The once-proud 100-percent-new-oak boast has been ditched as percentages drop. Barrel sizes are also getting bigger, thus reducing the ratio of internal surface area to volume, but most radically, new types of non-oak vessels are being used for fermentation and aging. Foremost among these are concrete eggs and giant amphorae, the latter used at Domaine de la Pousse d'Or in Burgundy since 2015 to vinify a proportion of the harvest from the premier cru En Caillerets vineyard. Tasted alongside the traditionally made wine from the same vineyard, the "cuvée amphora" is akin to listening to a musical recording with the bass notes turned down. The texture is lighter and the overall impression is one of delicacy rather than weight. The barrel makers must be feeling a chill wind of change, but there are new markets (China springs to mind) still to be serviced. The barrel's obituary is not about to be written.

The monster wines were in tune with their times, but their times are over. Eventually, they will be fondly remembered as warped wonders. History will judge them kindly—villains always are. Fickle fashion now lauds harmony and restraint but it never stands still. As sure as night follows day, the monsters will return. We are in a good moment now; enjoy it now.

8

ON LIGHTER FEET

The 2008 financial crash dealt a killer blow to all things gross, including heavyweight wines. At a stroke self-effacement replaced self-promotion, and the century's second decade started with a whimper. Change had been in the air prior to 2008, and the crash accelerated it, with wine styles mirroring the new world order. Talk of freshness abounded and a wine's texture, its tactile impact, featured increasingly in commentary. Previously, my own dalliance with the heavyweights, including a torrid engagement with Californian Zinfandel, had come to a close as I anticipated the move to slimmer styles.

Fashion's pendulum swung toward lightness and will continue its arc until stick-limbed wines of meager merit are celebrated. Currently, vibrant wines are in vogue and there will be more of them, slim yet toned, until they morph into alarming thinness, all sinew, no flesh. This new paradigm—severe and screeching—will be lauded to the heavens until a brave person calls them out and nudges the pendulum back toward the other extreme. Some white

Burgundies have already drifted to super slimness, with the oak and over-ripe fruit throttled back to expose the acidity, which emerged initially like a fresh breeze in a stuffy room but the breeze is now a gale. Somebody should close the window.

As belts were tightened post-crash the wine world woke up to subtlety, light and shade, tingle on the palate, not thump. And while this was happening it was business as usual in one of France's classic regions, where fleet flavors abound: the Loire Valley, one of the world's most overlooked wine regions. Perhaps its diversity, yielding a cornucopia of styles, counts against it. The world loves a simple message: Bordeaux hangs its hat on claret, despite the other styles produced there; ditto the Rhône, structured reds are its calling card. Burgundy's renown rests on a single grape for reds and another for whites. The Loire's message is not simple. It is a mosaic of many styles, yet legions of consumers know only one—Sancerre and, at a push, its cross-river sibling Pouilly-Fumé. The Loire is held captive by people's idea that it consists of these alone. Which is to ignore a world of wine that stretches from this pair in the east through Vouvray, Chinon, Saumur, and others all the way to Muscadet way out west, where the salty Atlantic air is captured in the briskly flavored wines, currently the world's most derided.

France's longest river traces a huge arc as it meanders more than 620 miles from the heart of France, first toward Paris and then the Atlantic. The Loire is a dawdler, spilling hither and thither, as if unsure what route to take. For much of its length it is bordered by vineyards and jeweled by châteaux. These are elaborate concoctions of turrets and whirligigs, wedding cakes in stone. Disney meets wine in the Loire Valley, best seen at Château du Nozet, a Magic Kingdom edifice that fronts the wine business of Baron Patrick de Ladoucette. Nozet and other dazzling châteaux compensate for the river's lack of spectacle. The Loire Valley itself claims none of the topographical drama that so enhances the Mosel, the Rhône, or the Douro.

While the river meanders, Loire wines don't dawdle; a dart of acidity keeps them linear and focused, giving family resemblance to

Muscadet

Poor Muscadet. If ever a wine deserved rehabilitation, this is it. When the world fell in love with ripe tropical fruit it fell out of love with Muscadet's snap. Good Muscadet that is, the poor stuff is barely more flavorsome than water and less enjoyable. The proper stuff is light and lean, not thin. A mineral bite runs through the fruit flavors like quicksilver, adding intensity and length, and making it the perfect partner for the abundant seafood harvested from the nearby Atlantic. There, food and wine should be matched without ceremony, heap the plates and charge the glasses.

The most baffling thing about Muscadet's fall from grace is that when consumer tastes swung in favor of light "lunchtime" wines it was ignored in favor of anodyne dross in the shape of numberless vapid rosés. The hordes rushed headlong for those bland potations while ignoring the brisk, iodine-etched delights of Muscadet, the perfect alfresco wine. There's no accounting for taste.

the wide variety of styles. It surges in some, is tamer in others, but is always there, like DNA in an extended family. On the surface the wines differ markedly but at a deeper level they share a common base off which each establishes its identity. And, as the acidity links the wines, the river links the widespread vineyards, binding them into common heritage. The wines are generally lighter in alcohol than equivalents from other French regions, and such lightness makes them versatile with food, complementing flavors rather than swamping them.

The red wines have lived long in the whites' shadow. This was once understandable, for the acidity that made the whites lively rendered the reds bitter, with a "green" flavor, more bilious than emerald, dominating the meager fruit. Today's reds have been

Great Grape: Cabernet Franc

Hail this "cab" while you still can, for it is about to have its moment. For decades it has looked on as its Sauvignon cousin conquered the world, renowned for its structure and depth. These qualities were not readily associated with Cabernet Franc, which is lighter on its feet with a character often described as "leafy"—sometimes a euphemism for raw-edged wines of limited appeal. There was a skittishness, a flavor inflection alien to the better-known Cabernet, like a garish tie worn with a pinstripe suit.

Until recently this acted as an impediment but now, under the twin influences of better viticultural practices and global warming, Cabernet Franc has lost its scrawniness. The firmer elements are now wrapped in ripe fruit; they no longer scratch the palate. And this model is being repeated across the globe by winemakers eager to try something new, while cleverly maintaining the Cabernet name recognition. From its Loire Valley heartland Cabernet Franc is like a sunny winter's day; there's brightness rather than warmth. The flavor oscillates from sun to shade in an intriguing interplay between opposites. Cabernet Franc is on a roll. Time to catch it.

to finishing school, as witness the appealing amalgam of earthy substance and persistent fruit now found in the wines of Chinon for example. Fashion has also swung the Loire's way, and Cabernet Franc may yet prove the region's trump card as its popularity surges around the globe. Loire reds are now lithe; what was dismissed as frail is now lauded as fresh, and the wine world's axis has shifted in the region's favor. In the wake of the monsters' hegemony Loire wines are like evening's cool after a sultry day.

However, it is understandable that the Loire never achieves the elite status bestowed on the twin powers of Bordeaux and Burgundy.

Great Grape: Chenin Blanc

In youth a great Chenin crackles on the palate—Chenin has never bothered with charm. It has some of the cussedness of Aligoté—a snappy wild child—akin to a brilliant-or-bad tennis ace, noted for glorious victories and memorable meltdowns. It can be coaxed to excellence but that twist of character remains; there is always a hint of danger. You come to appreciate and then love Chenin; it's unsettling at first, and thereafter enduringly brilliant. It's the bucking bronco that needs taming.

I speak of proper Chenin Blanc for, if ever a grape was yoked to the grind of overproduction, it was Chenin, most particularly in South Africa, where for decades it was beaten into blandness. Times change and South African winemakers are now cognizant of the quality they can wrest from Chenin yet, impressive and all as their wines are, it is in the Loire Valley that the grape reaches its apogee. There, it's also known as Pineau de la Loire and is appreciated for two trump cards: versatility and longevity. The latter marks it as possibly the longest lived of all grapes, a quality largely ignored by most wine drinkers. Chenin Blanc's place center stage has yet to come. Perhaps it never will.

Delightful as the wines are, they are seldom profound, seldom have the stop-you-in-your-tracks impact that signals peerless excellence. Most lack the indefinable "other" that sets the greatest styles apart from the very good and, much and all as I appreciate them, I am not blind to this appealing simplicity in place of complexity. They are a movement from a symphony, not the complete work. The Loire does many wines well but is short on paradigm styles that are imitated around the world.

An exception might be made for Vouvray, which, on the basis of its longevity can claim to be the Loire's greatest wine. It is also remarkably varied and can be bone dry, semi-sweet or lusciously so, as well as sparkling, yet until recently its charms lay hidden, known only to aficionados. Ironically, its age-ability can count against it for, in its ultra-dry incarnation, it is uncommunicative in youth. Sometimes decades are needed for it to blossom but riper recent vintages now give hope of more youthful appeal. There are fewer unripe youngsters that jangle on the palate.

Huet is the name to look for when seeking benchmark Vouvray: poised, accurately flavored, and virtually ageless, wines whose life spans are measured in decades. They have a nervy energy, active on the palate, not passive, expecting the taste buds to sit up and take note when they come visiting. They bend to no will and can be cussed, but when on song they are glorious. Idle, languid pleasure is not their forte; they engage and stimulate and challenge and only then do they reward the attentive drinker. Easy charm is for other wines.

The Loire does not have proprietary rights over lighter styles and the move toward elegance is allowing other long-established regions to reemerge, as an ebbing tide reveals a splendid beach. The wobbly wonders are struggling, like a middle-aged gent trying to fit into a suit from younger days. The forgotten classics are being rediscovered, none more so than Beaujolais, whose fall from public favor was a self-administered cataclysm. The trigger for this was Beaujolais *nouveau*, which raised the image while destroying the reputation. Everyone got to hear of Beaujolais—while also learning it was cheerful quaffing plonk and no more. As a two-edged sword, *nouveau* has few rivals in the history of wine.

The Beaujolais region, lying north of Lyon, traditionally sent a deluge of just-fermented wine—*nouveau*—into the city's casual restaurants, *les bouchons*, every November. It was served in jugs and cut a juicy swathe through the hearty cuisine. Things changed in the 1970s and '80s when the potential for worldwide distribution was exploited, a move that landed all Beaujolais in the cheap 'n' cheerful pigeonhole, from which escape has proved difficult. Unwittingly, the *nouveau* acted in tandem with the forces then ranging against lighter styles. Beaujolais's reputation was ransacked with the cooperation of the people who should have been protecting it. It was like leaving the front door of your house open to allow thieves easier access.

The *nouveau* wine is inoffensive, sometimes charming; its crime was the damage it did to the real stuff. Lips curl when I champion the joys of proper Beaujolais, and I struggle to cut through the disdain that is *nouveau's* child. It's a mystery that it still has proponents. People of otherwise sound judgment celebrate its flippant charms, seemingly unaware of the damage their advocacy does to good

---

Great Grape: Gamay

There was no equivocation when Duke Philip the Bold of Burgundy banned Gamay in 1395: "And this Gamay wine is of such a nature that it is extremely harmful to humans, people who drank it were infected with serious diseases because the wine which comes from the plant . . . is full of huge and horrible bitterness. . . . For this reason we order all who have Gamay vines to cut them down . . . within five months." Was ever a grape so damned as Gamay?

Perhaps a hard night on the Gamay prompted the duke's vitriol. Not many grapes boast such condemnation on their résumé and in the centuries since Gamay has been dogged by comparison with Burgundy's standard-bearer, Pinot Noir. It's a comparison that would shame most grapes, but only Gamay lives in close proximity to Pinot. It has made its home in Beaujolais, though it waves a defiant two fingers at the Côte d'Or from the tiny village called Gamay, beside Saint-Aubin and not far from patrician Chassagne-Montrachet. Some vines are planted there but it is in Beaujolais that it reigns unchallenged and from the best sites, the ten *crus*, it produces wines of real class that age well. Seek them out.

Beaujolais, which is glorious wine with a direct appeal, easy to understand, and splendidly quaffable.

For about two weeks every year *nouveau* raises Beaujolais's visibility long enough to confirm people's view that it's pleasant plonk. Perhaps it was a good idea once but it backfired in a way its proponents would never admit. Some may claim that consumers discover Beaujolais through *nouveau* but it is a dead end, not a slip road to greater delights. The ballyhoo of the past has died down but there is still too much fuss made of this hollow exercise. Every year my inbox fills with breathless messages signaling *nouveau's* arrival, apprising me of its delights and entreating me to favor it with a glowing recommendation. I decline and delete.

What's to be done? A partial solution is to hand, a collection of lyrical names unsullied by *nouveau*: Saint-Amour, Juliénas, Chénas, Moulin-à-Vent, Fleurie, Chiroubles, Morgon, Régnié, Brouilly, and Côte de Brouilly. These are the Beaujolais *crus* and they should be championed by the authorities. The debased Beaujolais name is seldom seen on the labels of these wines, the snag being that these are the higher-end offerings. Basic Beaujolais cannot be called Fleurie, for example. But for now they should be championed, allowing consumers to rediscover Beaujolais from the top down. They haven't done so from the bottom up. A new approach is needed—and the top wines are not hugely expensive. It's a bold suggestion, but boldness is needed to resuscitate the Beaujolais name.

Notwithstanding France's claims, the true home of light—not frail—wines is Germany, a nation more celebrated for beer by the liter than wine by the glass. Riesling is Germany's signature grape and it achieves excellence in many of her wine regions while in one it produces featherweight wines of transcendent beauty. They are as puzzling as they are satisfying, for it remains a mystery that something so delicate can still have the intensity and durability of a polished diamond. These are ballet dancers made wine. They hail from the Mosel, source of nectar to the cognoscenti but largely ignored by mainstream consumers. Such a situation seems inexplicable, though its parallel in the world of literature sheds light. Mosel Riesling is the poetry of the wine world; great poetry is the summit of literature but that does not convey popularity. We stick to prose, more easily accessed and understood.

The Mosel River writhes like a demented snake, twisting and turning, favoring one bank and then the other with prized south-facing slopes upon which the vines inch toward ripeness, needing every scrap of sunlight to get there. The steepest sites are best, tilting the vineyards toward the sun as well as the ribbon of river that reflects a precious ration of light. Tending the vines is a job

Great Grape: Riesling

Riesling—say "reeze-ling"—has suffered more than most grapes, a victim of fashion and clumsy legislators. Step into the dock the 1971 German wine law, the gist of which allowed Riesling to be blended with lesser grapes, with no limit set on yields. That over-simplification catches the nub of the problem—made manifest by the rampant 1970s popularity of flower water Liebfraumilch. That travesty masquerading as wine cut the legs from under the German quality wine industry and, by extension, any wine in a tall, graceful bottle. Riesling was damned.

Where Chardonnay, relatively speaking, offers a blank canvas for winemakers to put their stamp on, Riesling arrives with an outline sketch complete, in the form of tingling acidity. Plant it anywhere on the planet and, like an émigré, it might adopt local habits and customs but its accent never changes. It can shimmer and snap, as in Australia's Clare Valley, and it can caress and cuddle, as in its sweet Alsatian incarnation, but behind those widely differing flavors there is always an electric charge that marks Riesling as one of the greatest grapes of all. Perhaps the greatest.

more suited to mountain goats than human beings. Lung-bursting slopes are so steep in places that winemakers caution visitors to remain on the path adjacent to the vineyard while they step in to demonstrate the incline. Only the fleetest of foot are capable of the Herculean effort that working these vineyards calls for. And the slopes are treacherous not only because of the incline but also the loose scales of slate that skitter away underfoot, snapping ankles and twisting knees.

The Mosel scares away all but the hardiest souls, those for whom relentless labor holds no fear. They endure, sustained only by the shaky prospect of vinous magnificence. To wring such wonders from the Mosel's vineyards is a winemaking feat like no other. Yet their reward is meager, measured in rapture from a small band of devotees, while the majority look elsewhere. That something so delicate, so scented and succulent, so fragrant and fine can emanate from such conditions is remarkable; that it is then shoved to the margins is criminal.

The Mosel's two tributaries, the Saar and the Ruwer, are sources of equally compelling wines. Perhaps the most compelling of all come from the Saar, in the shape of Egon Müller's Scharzhofberger wines that set the gold standard for Riesling worldwide. A sip of these reveals a thrilling interplay, as the flavor dances between sweetness and acidity, with your palate as their playground. Müller and other top estates belong in a league of their own, able to command proper prices for their wines, while the less celebrated players soldier on, hopeful that a wine world besotted with drab homogeneity will someday wake to their charms. They have been ignored for decades, as wine lovers sought out more robust styles or, worse, wines with a total absence of character.

Great Mosel Riesling is the graphene of the wine world: wafer-thin yet durable as steel. The wines are rapiers to others' broadswords, with flavors so finely etched they escape capture in a tasting note. Riesling reaches a level of excellence in the Mosel that places it at the pinnacle of vinous achievement, equaled by some, exceeded by none. The wines bear testament to a near-mystical marriage of grape and ground; no other combination of region and Riesling comes close to replicating the Mosel. Trying to identify its essential character is not easy, but it achieves an almost weightless quality on the palate, and with many consumers seeking out lower-alcohol wines, perhaps they will discover the Mosel. Perhaps.

It is not only long-forgotten traditional styles that are having their moment in the sun, as the popularity of palate thumpers declines. The natural wine movement chimes with the "lighter feet" zeitgeist and sprang up as a response to mass-produced wines, those that are manipulated and subjected to inputs and additives at every stage, from vine to bottle. Few wine drinkers know the large-scale industrial practices employed to turn out reliable wines at low prices. As a raw material, grapes come with challenges: weather vagaries influence quality and, more distressingly for the accountants, quantity. That upsets growth forecasts and profit projections, so vines must be planted in places where they ripen reliably and bountifully.

Not many have seen the articulated trucks of grapes arriving at the winery to deposit their multi-tonne loads—leavened with leaves, twigs, insects, bird droppings, and other non-grape detritus—for processing into wine. That process is rigidly controlled to yield a product of impeccable credentials when subject to laboratory analysis. It is faultless in the narrowest sense, meeting preordained criteria and ticking boxes in robotic fashion; and satisfying in the most basic sense, like a catchy advertising jingle. Natural wines reacted against production methods that wrung the life out of wine, methods that drove them rigidly down the middle of the road. In style, industrial wines are made not to please all palates but to offend none; they are the magnolia paint of the wine world, eschewing diversity, wine's trump card, and embracing homogeneity, its greatest enemy.

Natural wines fit the less-is-more template, with a side order of funk. But "natural" defies definition; it is a woolly term, open to abuse and misinterpretation. It is bandied about, especially by those selling the wines, who exploit consumers' trust of the word. "Natural" preys on our broader use of the word, never questioning its meaning in whatever context it is used, simply relying on the cozy glow it engenders. We all know the meaning of "natural"— until asked to define it. Outside the wine world it is used with glib abandon, invoking images of unadulterated purity. Something un-messed with, something that can be trusted.

In that context its meaning is close to "not man-made," so food may contain natural coloring or flavoring that was derived from something that grew in a field rather than a lab. But that meaning doesn't fit natural wines, which are man-made, and no amount of hands-off winemaking escapes that fact. The vines don't plant or prune themselves, the grapes don't harvest themselves, jump into the vat, and ferment away in jolly old fashion. Nor do they bottle themselves or sell themselves. They may be less manipulated, with fewer additives than conventional wines, but their production involves plenty of hands-on attention. And does shipping them around the world not introduce a fly in the ointment? Does that chime with their makers' concern for the planet?

Our trust of "natural" is bad for natural wine. Some scrutiny might be painful but would also force socks to be pulled up and stop people hiding behind assertions of vinous rectitude. More rigor and less exploiting the vagueness of the term would be welcome. It is understandable that winemakers piggyback on its appeal, if only because the alternatives are clodhopping: unadulterated, non-interventionist, low intervention. The latter is more accurate, but natural wins the day for simplicity and the attraction of allowing consumers to put their own meaning on it. They are rated more for their right-on production philosophy, than for their obtuse flavors. But this is a cart-before-horse way to assess a product designed for sensory pleasure. Flavor is all, and all wines must be judged on the palate. If this was done it would act as a stimulus to the producers; giving them a free pass will lull winemakers into laziness, convince them they can rely on woolly feel-good assertions rather than quality to achieve success.

Most natural wines have an appealing rawness, they boast a purer expression of what the grape is capable, and they are not coiffed and primped. Initially the taste buds respond well to their uncomplicated charm, but sustained attraction is another matter. Trade tastings featuring a slew of them can be wearisome and often run to the soundtrack of exultant murmurings from their fans. I once followed a disciple as he made his way around a tasting, purposely sampling all the wines that sent him into raptures, and was left un-raptured. Some did have a vivid charge of flavor but some were crude, function-over-form wines like the work of a blacksmith, not a silversmith. Suggesting the rustic charm of the former is as elevated as the refined elegance of the latter is misguided. The obvious hierarchy should not be ignored; I want a silversmith's wine, not a blacksmith's.

Discussion with natural wine advocates can be challenging, they have bought into a movement fueled by dogma. Too many are dazzled by the correctness of their beliefs, unaware, or unwilling to accept, there are many paths to vinous nirvana. Perhaps the crux is that I see the wines as different, while they see them as better by definition. They have seized the moral high ground, which can be a chilly perch if you don't follow through on your initial promise. They claim a vital energy for their stripped back wines, a flavor surge untrammeled by the manipulations necessary for industrial-scale production. Nonbelievers might label it an untamed element that can be feral at times, and wonder if we should leave everything undeveloped, unrefined? As humans we are more natural if we don't comb our hair or wash but do we look our best first thing in the morning? Should we go back to bearskins and body odor? Refinement, taken too far, warps the original, but not done at all is like leaving a diamond in its raw state; it needs cutting and polishing to reveal its charms.

Yet I also fall prey to judging natural wines by a standard other than flavor. I applaud them because they are giving the wine world a jolt, shaking complacency and challenging practices set in stone for too long. The natural wine movement deserves credit as much for its disruptive influence as for the new styles it has introduced. They should be lauded to the heavens for ruffling feathers. If only a scrap more character ends up in bland brands, then natural wines will have been a success. Additionally, they must be celebrated for helping to preserve wine's variety and diversity—the two pillars that put it ahead of all other beverages. Wine tells a story like no other drink, yet if its makers treat it as simply another beverage, another alcoholic liquid to be primped and packaged and marketed and sold at a set price point, then it is doomed to languish in the bargain basement aisles of supermarkets and discount stores, stripped of any compelling reason to buy this wine or that, apart from price.

Natural wines play the diversity card with gusto, and I salute their proponents' individualism, but the dogma that attends their pronouncements gives their critics plenty of ammunition. Opprobrium has been heaped on them and though their response can be prickly, I suspect them of secret delight, for it also acts as validation. The critics frequently miss the point: if you want characterful wines, then you have to take the rough with the smooth. Wines that forsake the middle of the road are less consistent and more likely to divide opinion, a circumstance to be welcomed. Maybe natural wines won't endure, like a pacemaker in a race who does his job then slips away. Nevertheless, they cried halt and, like them or loathe them, they performed this service on behalf of every wine lover who cherishes a decent glass with character and identity. Thus they are more appreciated than liked and, although they walk on lighter feet, they are an adjunct to the move toward lighter styles rather than a main player. They have challenged lazy norms, and if they don't endure they will be looked back on as the punk rockers of wine. Their epitaph might be: "They shook things up."

9

ACCESSORIZE

Are wine lovers more gullible than other collector types? The proliferation of gadgets and gizmos that promise enhanced pleasure from every bottle suggests "yes." Wine attracts accessories like a flame attracts moths. Where there is a wine geek there is a clever marketer waiting to convince them there is an accessory they want, regardless of need. Most end up thrown in kitchen drawers or lost at the back of cupboards. Many of them fall into the one-use-wonder category: used initially with great excitement and then discarded because they didn't deliver on the manufacturer's promise. Some are ill-disguised attempts to leech money out of wine drinkers—or their loved ones, desperate to buy them a gift. Too often they involve a tedious procedure that calls attention to itself. An accessory should be just that, something that sits on the margins, amplifying enjoyment but not distracting from the wine.

The accessory spectrum runs from cannot-do-without to useless. At this latter end fortunes have been made by duping people into purchases, many of which end up on the regifting circuit. Pride of place goes to the cellar book, in which one's wine collection is recorded and tracked. These impressive tomes were popular in the 1990s, usually gifted at Christmas and lovingly filled in by the recipient in the late December doldrums, before abandonment in the new year. Rediscovery some months later brought frustration; with two pages filled in it couldn't be regifted? Or could they be excised with a razor blade? At the cannot-do-without end of the spectrum there are two essentials: corkscrews and glasses.

The corkscrew is the daddy of all wine accessories. English: Corkscrew. French: Tire-bouchon. Italian: Cavatappi. German: Korkenzieher. Spanish: Sacacorchos. Greek: Tirmpousó (Τιρμπουσό). Hungarian: Dugóhúzó. Portuguese: Saca rolhas. Russian: Shtopor (Штопор) . . . Whatever you call it, however you spell it, the corkscrew has been with us for centuries, an essential accessory whose importance is barely diminished by the rise of the screw cap.

Great ingenuity has been brought to the challenge of reliably removing a cork from a bottle. The earliest corkscrews were simple: a metal worm with a handle in a "T" shape that was twisted down into the cork to gain purchase and then pulled out. That original is still with us, joined over the years by a bewildering array of others—some ingenious, some not—all of which use a form of screw or leverage mechanism to ease the task. Some designers reveled in complication, adding whimsical ratchets, levers, and screws of little utility. Most corkscrews have remained simple and unadorned but some are gorgeously decorated, almost like pieces of jewelry, with precious metals, rare woods, stones, and artwork incorporated into their design.

Some collectors have amassed thousands of them—I once spent a day inspecting a private collection, cataloged and cross-referenced, each one tagged and logged with date of purchase, price paid, and other relevant details, the ones in undamaged original boxes being the most valuable. There are notable public collections

Champagne on Tap

Time was when the metal disc on top of the classic Champagne cork had a hole in it to facilitate the insertion of a Champagne tap. This allowed for small amounts to be drawn from the bottle without it being opened. Which begs the question—why would anyone ever want a small quantity of Champagne?

My tap consists of a thin, hollow tube, threaded to a point, which allows it to be screwed down through hole and cork. The protruding end is opened and closed by turning a tap that allows the Champagne to flow through the flared opening. It is date stamped "Law & Co 1917" and I have never used it, though there's a bottle of Mumm 1952 in the cellar, bought at auction a few decades ago, with the requisite hole in the top.

also and one of the finest is found in the Vivanco Museum of Wine Culture in Briones, Rioja, Spain. The numbers are overwhelming: scores and scores of corkscrews in display case after display case, each containing up to one hundred models in row upon row, some only minutely differentiated from their neighbors.

It is easy to be dazzled by these, but we shouldn't lose sight of the corkscrew's purpose. The best ones, such as the Forge de Laguiole, involve no fuss and bother, and it is worth investing in one. I have had mine for over twenty years and it works as well now as it did with the first bottle—a tribute to great design and excellent engineering. Laguioles are expensive but are also of great value, for they are almost indestructible. The caveat is that you should only splash out on one if you are the sort of person who doesn't lose such items. And have a backup one for away fixtures such as picnics, especially if they are beside water. My Laguiole never leaves the house.

The best backup is the "waiter's friend," the type of corkscrew that many sommeliers use, which works very well and fits easily into a pocket. The only drawback is that after prolonged use the joints go a bit arthritic and then it is time to replace. Another good one is the Screwpull, a clever beauty that comes in a variety of colors. Unlike most corkscrews this works by "lifting" the cork on a continuously turning worm, useful for anybody who does not have great strength in their hands. Perhaps the most clever "corkscrew" of them all is the "Ah-So," which, strictly speaking, is not a corkscrew at all, and is more accurately described as a "cork remover" or "wine opener." It consists of two thin, asymmetric steel prongs attached to a handle. The longer prong is inserted between the cork and the bottle and pushed down until the shorter one is inserted on the opposite side of the cork. Using a rocking motion the two prongs are then gradually inserted to their full length. A twisting motion then removes the cork easily. The Ah-So is particularly good at removing old, fragile corks that are wont to crumble if a conventional corkscrew is used.

I am a sucker for wineglasses, and, judging by the endless stream of new collections and styles, I am not the only one. Glasses range from perfect-for-purpose to whimsical, the most ludicrous being the Champagne glass where stem and bowl are one, hollow from base to top. It's a glorified tube that flares outward, like an upright trumpet. Unless you clutch it by the base the heat from your fingers warms the wine, rendering it flat and charmless. If I set out to design a glass to ruin a wine yet have great novelty appeal, I could not have done better.

The business of getting wine from bottle to palate is big, and growing at a pace that is difficult to keep up with. So similar are many of the modern glass types and styles they need to be examined side by side to discern any differences. What they all share is super slimness, phenomenally light yet not overly fragile. And their proportions are less gross than previously, when the biggest glasses, led by the Riedel "Sommelier" Burgundy and Bordeaux grand cru pair, dominated a dining table, great bollards at constant risk of toppling. They were once cutting edge but when spotted today look like something from a freak show, as classy as a stretch limo filled with stag or hen party revelers.

A friend once quipped that they were "pint glasses on stems." The only thing he got wrong was the capacity—these were liter glasses on stems. A full liter could be squeezed in, meaning the bowl could easily hold a standard 75 cl bottle in one pour, perfect for anyone abiding by one-glass-a-day medical advice. A reasonable pour formed a puddle in the glass. My experience with them was that all but the most powerful wines disappeared into their depths— most memorably, a 1961 Château Lafite that struggled to fill the glass cavern with any aroma.

But credit where it is due, it is thanks to Riedel from Austria that wine lovers have such choice and quality in glasses. They rewrote

Great Grape: Blaufränkisch

Blaufränkisch goes by many names, collecting them like stamps on a passport: Kékfrankos in Hungary, Lemberger in Germany, Modra Frankinja in Slovenia, but it is as Blaufränkisch in Austria that it shines brightest. There, it calls to mind a blend of Loire Cabernet Franc, northern Rhône Syrah, and Côte de Beaune Pinot Noir, yielding an intriguing witches' brew that oscillates between these poles, defying attempts to pin it down. Perhaps that difficulty explains its lack of recognition internationally, for it has yet to garner the sort of acceptance that its quality warrants.

It is Austria's secret weapon, still dismissed by some as an also-ran grape, never accorded a seat at the top table. Might these be the same people who for years were immune to Grüner Veltliner's pedigree? Blaufränkisch is now Austria's leading red grape, but it may need a few years before it is crowned as such, sweeping aside the likes of Zweigelt or St-Laurent. Nevertheless, its chiseled and precise palate—nervy and tense with a hint of danger—may keep it as a niche player. What a niche.

the playbook, almost single-handedly moving the world away from the all-purpose Paris goblet, the clunky vessel that served for all wines and occasions and resembled a glass tennis ball on a stem. The Riedels, father and son, Georg and Maximilian, are constant innovators and master marketeers and they preach their gospel with messianic fervor at stage-managed masterclasses. These "prove" how each glass is uniquely suited to the requirements of specific wine styles.

The events are solemn and run with Teutonic precision, and no deviation from a preordained path is tolerated. Attendees are watched with a gimlet eye to see they toe the line. On one occasion myself and a colleague were sitting well back in a crowded room, when he idly opened the individual bottle of mineral water provided for each attendee. A sharp rebuke from the stage brought him up short. Water would be drunk to instruction and not indulged on a whim. I added a mock slap on the wrist and we all chuckled— sometimes these events are only made bearable by the unintended comedy.

Great belief in one's product is needed to bend a room full of people to your message and the Riedels, performing as circus ringmasters, do it with endless prompts in near-clinical conditions. In such a situation, where the same wine might be tasted in several different glasses, there is a perceptible difference to how it tastes from each. But when I am at home and perhaps use the "wrong" glass for the style of wine I am drinking I don't feel my enjoyment is greatly diminished. And that is the true test, ordinary conditions, rather than contrived settings where almost anything can be demonstrated. The slick performances are compelling initially but, like a conjuring trick seen too often, they lose their sparkle; there's only so many times you can be cajoled toward belief in the glasses' efficacy.

The Riedels blazed a trail and many followed in their slipstream, all producing ever more refined shapes that threaten to outshine the originals. Riedel responded with eye-catching creations, though two recent ranges make me wonder if the creative barrel is being scraped. The faceted Performance glasses, with tiny inflections on the bowl only perceived when the glass is rotated, are conversation pieces, yet will we ask in time: "Do you remember those quirky glasses?" Similarly, the machine-made Winewings range has a "wobble" in the bowl. If it was mouth-blown we would wonder if the blower had coughed mid-puff. Neither is an improvement, merely an addition to an already overflowing catalog. The previous Veritas range trumps them both and comprises some of the most beautiful glasses the Riedels have ever made.

If Riedel's sensual curves cast them as the Ferraris of the glass world then archrival Zalto are the Lamborghinis, all arresting lines and sharp angles. From the pencil-slim stems to the wafer-thin bowls, these make a visual statement like no other. They were designed with input from Austrian wine expert Father Hans Denk, hence the base of every glass is etched with "Denk'Art." As with every manufacturer today the sales patter comprises plenty of earnest techno-babble, and in Zalto's case this takes the form of explaining that the glasses' sides are angled at 24°, 48° and 72°, which somehow

Stemless

Suitcases have handles for a purpose and wineglasses have stems for the same purpose—for holding them by. Tucking a case underarm is awkward, just as holding a glass by the bowl is. Hold it by the stem, which is what it's for. And it is not simply a handle; it allows visual appreciation and stops clammy palms from warming the wine. A plump bowl held aloft by a pencil-slim stem is aesthetically pleasing too. Stemless wineglasses have recently made an appearance and they are great—for water. Otherwise, stemless is dumpy, the proportions are wrong, and they look like a person lopped off at the waist. Can a stemless glass even be called a wineglass? Not by me.

puts them in harmony with, "the tilt angles of the Earth," according to the company's website. One's enjoyment of wine is enhanced, "due to these cosmic parallels." Not even weapons-grade guff such as this can hinder their appeal, which was underscored a few years ago when the London wine club, 67 Pall Mall, caused a stir by ordering nine thousand Zaltos, an order of such magnitude that the factory was closed to other orders for six weeks.

Other glassmakers on the merry-go-round include Sydonious, Grassl, and Lehmann, all vying to concoct difficult-to-dispute arguments in favor of their glasses as they strive for market share. Talk of critical dimensions and curves and radii abounds, dispensed like trade secrets and described in detail to underpin the claims and sway consumers in their direction. A mumbo jumbo detox is needed. In all this the manufacturers seem to forget that enjoyment is the key to wine, not cerebral assessment. Joy is being swamped by cart-before-horse clinical thought. Wine is a sensual pleasure, and the glass should aid that, rather than claiming to be a laboratory on a stem.

Another type of claim entered the stemware arena recently when noted wine writer Jancis Robinson launched a "range" comprised of a single glass, said to suit every style of wine. It's a beauty, graceful and so fine it looks like it might levitate. The sibling decanter for young wines, on the other hand, is heavy-hipped and is clunky compared to the glass. It is hard to believe they both came from the same design team. While I love the glass, I am not won over by the thinking behind it; it's too utilitarian. I like a variety of glasses and, when serving wines side by side, different glasses make for easy identification, which is useful as a bibulous dinner takes flight. Nor am I a fan of the other extreme where the proliferation of styles is impossible to keep up with. A core range of about half a dozen is the sweet spot, with extras for Champagne and fortified wines.

One common attribute of modern glasses is that they are dishwasher-friendly, something that many people still fail to grasp as they opt for hand-washing and more breakages than would ever happen in a dishwasher. The trick is to use a short cycle, take them out immediately, and give them a quick polish. Don't put them in last thing at night, leaving them to stew while you sleep. There might be the odd breakage and it is only when you collect the featherweight wafers of glass that you realize how fine they are and how sophisticated the manufacturing process is.

As glasses have gotten lighter, curiously, wine bottles have bulked up, piling on the weight, and are now plus size to glasses' size zero. Sitting between them are decanters, some looking like pieces of sculpture, riots of sinuous beauty and complete impracticality. Some writhe like snakes, and it takes a conjuror's hand to pour from them. Cleaning them is a nightmare. I confess to a soft spot for decanters, though it has hardened over the years as the tedium of washing

Cleaning Decanters

"Is it a toilet brush?" asked the wife as her husband set about cleaning a stained decanter with a long, flexible implement festooned with bristles and odd spongey bits. He didn't see the humor—cleaning decanters is not fun. The best advice is to clean them straight after use, which is about as helpful as saying abstinence is the best cure for a hangover.

Apart from giraffe-like brushes, there's an impressive arsenal of devices and implements whose purpose is to remove that murky stain smiling at you from the bottom of your favorite decanter. Denture cleaner is favored by some but there are occasions when a more abrasive approach is needed, which is where "Magic Balls" come in. Swirl these tiny ball bearings, perhaps one hundred of them, with some water to scour the stain without damaging the glass. Uncooked grains of rice perform the same service, as does sand, though it's a bit messy. During all this you can marvel at how stubborn the stains are— and promise never to let them intrude again.

them challenges the joy of using them. Now I use the simplest ones, to invest the service of a special wine with some ceremony, and leave the sculptural beauties on the shelf for distant admiration.

Corkscrews and glasses date back centuries, but the accessory du jour is a recent invention. The Coravin is barely a decade old, yet it is already established as an indispensable tool for professionals and for some keen amateurs too. Put simply, this device has solved the age-old problem of how to serve a glass of wine from a bottle without opening it. It works by inserting a slim, hollow needle through the cork and, as the wine is extracted, inert argon gas flows back in to make good the deficit.

As soon as it was launched people fell over themselves to applaud it so, when given one to test, the skeptic in me chose a more considered approach. Taking three bottles of different wines I served two glasses out of each using the Coravin. A week later I opened bottle one and found it to be unchanged. A month later I opened bottle two with the same result. Bottle three followed eleven months later and tasted as mediocre then as it did on first tasting. So, full marks, though I couldn't help quipping that what I wanted for Christmas was a Coravin device that could improve mediocre wines. That would fly off the shelves.

Once convinced, however, I never used it again. It wasn't a conscious decision, it just fell out of use. Coravin is the daddy of all gadgets. In actuality, describing it as such does it a disservice, for there is nothing gimmicky about it, it is an excellent device, though it doesn't fall into the cannot-do-without category beside corkscrews and glasses. The only drawback for the amateur is that it isn't cheap and frequent use runs up more cost in replacement gas cartridges.

Seeing it in action, in a restaurant setting for instance, reveals an inelegant operation as a ration of wine is expelled, then replaced by the argon, then expelled and replaced. Even knowing its efficacy, I recoil when it is used to pour a glass at table. It's too clunky. Its stomping ground is the professional arena, where it allows a wine trade rep to use a single bottle of a prestige wine to present to sommeliers in different restaurants for tasting, or a winemaker can pour samples over an extended period of time to a series of visitors. Amateurs should save their money, safe in the knowledge that an open bottle keeps for much longer than is generally realized.

Every wine lover hankers after the ultimate accessory—a classic cellar whose ancient stone arches were crafted centuries ago, with dangling cobwebs and gloom held at bay by guttering candles. Most have to make do with an under-the-stairs wine cubby or a simple rack in a corner of a spare bedroom or a disused wardrobe. Even those whose pockets run deep seldom build a classic cellar to store their precious bottles, opting instead to indulge in glitz and glam, resulting in extravagant wine temples whose contents are venerated with quasi-religious fervor.

In these, racking is artfully crafted, with customized niches for large-format bottles, lighting is subtle, and various *objets*—antique corkscrews, old corks, labels, and empty bottles—vie for the visitor's attention. There may even be a dining area, with a table made from repurposed vats. It is sterile and inspires awe, though perhaps not envy, for it is little more sophisticated than the naff home bars favored by drug barons in TV dramas.

The dubious magnificence of these vanity projects places wine center stage, but the drama is all around it. Weirdly, it struggles for attention as a bit player might. It is worshipped as great paintings are, safe behind a protective sheet of glass, but a bottle of wine is not a work of art—it must be destroyed to be enjoyed. Destruction and enjoyment are inseparable. It is easy to become too reverential about wine, to lock it away like a mummy in a mausoleum, to forget that winemakers want their wines to be drunk, not worshipped. Wine should be stored properly, but eventually the coup de grâce via corkscrew, decanter, and glass must be administered. Wine's greatest joy is to bring people together in convivial circumstances;

Bizarre Consequences

Extremes of temperature have the most dramatic effects on a bottle of wine. I have seen a noble German Riesling turned into a mushy pulp by being consigned to a freezer and forgotten for a couple of hours. At the other extreme, a Corton-Charlemagne that baked in the trunk of a car on a sweltering day expanded to the point where the cork was pushed halfway out of the bottle. Remarkably, after being given a few hours to recover, the Riesling drank very well. As did the Corton-Charlemagne, after the cork was pushed back down and it was rested for three months to recover from its trauma. A Champagne bottle left in a freezer too long will eventually explode, because the cork is held in place. Beware.

only by drinking it is the circle, started when the winemaker planted the vines, finally closed.

I look after my wine well but I don't venerate it. Some bottles have been with me for decades and have come through several house moves and varied storage conditions unscathed. One example was a bottle of Cloudy Bay Sauvignon Blanc 1997, received as a gift in December 1997 and only opened in April 2020. It was magnificent: buttercup yellow in the glass, tropical fruit on the nose followed by a plump, satisfying palate. More remarkable still, the half-finished bottle tasted superb twenty-four hours later, with a racier flavor, more in tune with Sauvignon's character. A similar story attaches to a Domaine Huet Vouvray, Clos du Bourg sec 1971, bought in the late 1980s and opened to celebrate a fiftieth wedding anniversary in 2021. It was delicious, as I noted in a social media post: "Didn't disappoint & the happy couple loved it. Gorgeous gold color. Old preserved lemon flavors. Showing some age at the edges but in a 'valid' fashion. Everything revolved around the shimmering core of acidity. Singing length."

My wines are now stored in a well-insulated, purpose-built space and experience gentle seasonal temperature fluctuations with no perceptible harm. They are not held in a temperature- and humidity-controlled straitjacket, which is best if you are planning to keep your wine for a century, but who does that? To some ears this is heresy, yet careful observation and long experience has taught me that it works for at least a couple of decades. The key is to avoid temperature extremes and sudden fluctuations, neither of which are characteristic of my Irish homeland. And if seeking advice on the subject, try to ensure it is independent of sales patter.

It is not an accessory in the strictest sense, but the tasting note is now so embedded in the fabric of the wine business that wines appear naked without them. The note, and attendant score, defend against the chill winds of consumer criticism. Weirdly, the more revered the wine, the more it needs tasting note and score validation. Some critics hand them down like tablets of stone and some consumers treat them similarly.

Wine writing that consists of tasting notes and little else is a desiccated prose that has held the wine world in its grip for too long. It rewards wine expertise ahead of writing facility. It sprang from taking the glorious liquid too seriously, examining it through a metaphorical microscope, seeing the trees while never glimpsing the forest. That approach begets dreary writing, form sacrificed on the altar of box-ticking function. It is correct in only the most dismal sense; it informs but doesn't engage or entertain. The formal study of wine has something to answer for here, because it casts imagination aside. Yet our appreciation of music or literature would be nothing without imagination. Why not wine? Imagination fires the intellect as yeast does bread and wine. If the writer holds the wine at arm's length, unengaged, the reader will not be engaged either. The tasting note should be a pen portrait, not a résumé.

It is baffling that so many critics shackle wine with lumpen words in an effort to convey its joys to others. It is the equivalent of pinning a butterfly down, the better to study it. There is no intrigue when it is pinned to a board. Without some mystery, there will be no enduring

Great Grape: Malbec

The world has been simultaneously kind and unkind to Malbec, on one hand granting it cult status, while on the other attaching narrow strictures to that status, demanding it produce wines that conform to a mono-dimensional interpretation of the grape's qualities. Malbec is cast as the grape that delivers a lumbering mouthful that lacks for nothing in weight while abandoning any pretense at subtlety. As a consequence Malbec is a seen as a heavy-shod wine that barrels across the palate, punching to left and right, trampling the taste buds. Perhaps.

Today, in its adopted homeland of Argentina, Malbec is slimming, swapping clout for elegance and a textural freshness, obtained by planting the vines at higher altitudes, where the cool nights help retain acidity. In this metamorphosis Argentinean winemakers are fortunate to have the towering Andes in their back garden. And tower they do: a vineyard at one thousand meters altitude is unremarkable, two thousand meters is getting high, and there are some in the far north above three thousand meters. That's high, and the result is wines with a more clearly etched flavor, intense rather than hefty and capable of delivering ripe fruit leavened by a savory tingle, with perky tannins kicking into a long finish. Malbec? The best is yet to come.

attraction. It is the same with wine. If a theater critic listed the cast of a play and commented on the sets, costumes, and lighting—and made no other comment—we would feel pretty let down. Some tasting notes veer close to that slack template. Wine is all about people and place, about stories. Why not tell them by painting a verbal picture? Thankfully that is now happening, and this accessory is having a makeover, being rescued from tinder-dry hands and placed into those with flair and imagination. Not before time.

* * *

The Drop Stop

White tablecloths and red wine seem drawn by magnetic attraction, the dark stains standing in accusation of an unsteady hand. No more. The "Drop Stop" is the small silver flexible disc that you roll into a cylinder and insert into the neck of the bottle to form an elegant pourer that never drips, ever. A recent modification sees a notch cut in the side to prevent it sliding down into the bottle. It can be branded. It can be washed. It can be used again and again. Its brilliant simplicity combined with flawless performance time and time again pushes it to the top of the accessory pile. Simplest is always best, and the "Drop Stop" is simplest of all.

"Can I manage without it?" is the acid test that should be applied to all accessories, and the only two that unfailingly come up smiling are corkscrews and glasses. Thereafter, we move across a spectrum of increasing promise but diminishing utility. The wine aerator sits at the promise/utility intersection. Wine is agitated as it passes through the device and is exposed to more air than if poured conventionally. The aerator works and is eye-catching as a pre-dinner cabaret, but the same effect is gained by pouring the wine into a large water jug and shaking it vigorously, as I once saw Jean-Claude Ramonet do with a bottle of his Montrachet. I regularly upend a bottle of young white Burgundy into a decanter, letting it splash and sparkle, foam and froth. Money spent on an aerator is better spent on wine.

Similarly, it is hard to get excited about the pouring cradle, an elaborate contraption into which a venerable bottle, typically red Bordeaux, is placed. Once opened, it is gently tilted by means of a screw mechanism which is turned slowly to decant the wine without disturbing the sediment. A steady hand serves as well. One home-spun "accessory" is the teaspoon inserted into the neck of a half-finished bottle of Champagne to prevent the remainder from going flat. This ranks with the best of urban myths and, despite being repeatedly trampled by the evidence, it refuses to die. Its greatest use is as a stimulant to flagging dinner party conversation, when it can be lobbed in as gospel to incredulous guests, but the question must be asked: Why not finish the bottle?

The glass-writing pen sits firmly at the utility end of the spectrum, rubbing shoulders with the simple Drop Stop, though with narrower appeal. It can be used to mark the base of glasses to identify different wines in a home tasting, and also to write salient information on bottles, such as date purchased or, if a gift, the name of the giver. It is not indispensable, but given that it costs very little it is a harmless indulgence that enhances enjoyment for those who like to squirrel away bottles for years before opening.

The message is clear: don't be dazzled by accessories that make outlandish claims. The greater the claim the greater the disappointment if it fails to deliver. Spend your money on essentials, remembering that wine is the essential of all essentials.

10

AND TO EAT?

Is there a wonder wine that matches whatever is on the plate? A wine that can bend to the breeze, flex and twist, ride with every forkful and still come up smiling? No, but there is one that comes close.

The outstanding candidate for consideration as a wonder wine is Fino Sherry, and its cousin Manzanilla, both of which are able to cope with just about everything, apart from luscious desserts. A memorable lunch years ago in Jerez, home of Sherry, underscored this versatility, when a single wine was served to accompany a series of dishes, starting with a bowl of olives followed by a plate of melt-in-the-mouth Iberico ham reared in open fields on a diet of acorns. Then a dish of outrageously flavored tomatoes drizzled with oil, sprinkled with salt, and suffused with garlic, before we moved on to a sturdier challenge in the shape of wild rabbit liver. Then *tagarnina,* a local delicacy of cardoon stems and eggs, followed by marvelous goat's cutlets.

My host and I shared a bottle of Fino and never was a more persuasive argument made for its versatility as a food match. It wasn't brilliant with everything, but other wines would have been rooted to the spot whereas Fino has a gymnast's flexibility when confronted by a series of dishes. Its pungent savor and backbone of alcohol act like stabilizers, keeping the wine on course despite the buffeting. More celebrated wines sing glorious duets when perfectly paired—think claret with lamb—but they falter when asked to be more adaptable. And how is this wonder wine treated? Despite being the decathlete of wines, it is ignored by huge swathes of the wine drinking world.

Notwithstanding Fino's versatility, trial by multicourse meal doesn't allow it to shine the way it does when paired with one carefully matched foodstuff. Many spring to mind: olives, toasted almonds, smoked salmon . . . but the best Iberico ham stands supreme. Serve the Fino as you would any white wine, chilled and in a proper wineglass, with the Iberico, a contender for the world's finest foodstuff, sliced wafer-thin. Then revel in the interplay between food and wine, as sublime and satisfying as that between violin and piano in a Beethoven sonata.

Fino's starring role is as the aperitif, where its costar is Champagne. The aperitif is the diner's warm-up, its purpose to get everything focused on the challenge ahead, stimulating and engaging the senses. Hence, I never fail to be baffled in Bordeaux when a choice of Champagne or Sauternes is offered as an aperitif— most memorably before lunch at Château Lafite. Sweet wine has a soporific effect on the gastric apparatus, dropping heavy into the empty stomach, falsely filling the void, whereas the bracing acidity of Champagne and the urgent rush of bubbles act as a siren call to action. It baffles me that a nation so attuned to the sensual joys of wining and dining should indulge in this travesty. It is like an athlete taking a sedative before a race. Another wildly popular aperitif in France is white Port. Vast quantities are sluiced back by otherwise sensible diners before taking their places at the table, which must please the Port producers immensely, for it is the humblest of their products, and they ship oceans of it to a country that should know better.

The business of matching food and wine can be a heated topic, with some fanatics espousing a forensic approach, examining the components of a dish to winkle out its flavor DNA, then linking it to a wine that answers the call. Taken to this extreme food and wine matching is a dismal exercise in nerdy one-upmanship. Its disciples strive to outdo each other, first in identifying a barely perceptible flavor and then in suggesting an off-the-wall wine match only they have heard of. It's a knowledge parade of little utility and dull amusement. It would be like running a dating service that matches up people with the same blood type. Such a convoluted approach withers my enthusiasm.

At the other extreme lies the "anything goes" brigade, composed of knowledgeable wine people who, when asked for food and wine

Surf 'n' Turf

Surf 'n' turf is in the dock, accused of being a lunatic collision of foodstuffs. Rhyme is its only attraction. It should be called out for the mishmash it is instead of being served up by talented chefs to approbation all round. It's a mess on the plate. How is it that this culinary catastrophe passes the juries of public and professional opinion unscathed? Through the course of a meal I'll likely drink white wine followed by red, but I do so separately. I don't mix them to create a pink concoction. Why do we do so with seafood and meat? I can just about tolerate scallops and bacon. But that's enough. The question of what wine to drink with this clash of sea and land is a thorny one. Don't punish a great bottle; stick with something simple and direct, and save your patrician wines for more judicious matches.

matching advice, fall over themselves not to be intimidating. They espouse a whatever-you-like approach; every whim and opinion is of equal merit. The neophyte may not be intimidated but is left confused, rudderless in an ocean of competing choices, the sham inclusivity of little use to them. They must be encouraged, but this is not the way to do it. Thus when someone asks me for wine advice, I try to give it and not fob them off with a "whatever" cop-out. If that was my attitude I might as well advise them to match steak and Sauternes. That pair alliterate, they don't match.

When I ask for advice on a subject, that's what I expect, not anodyne drivel, a verbal hand pat assuring me my opinion is as valid as any other. A valid opinion is based on knowledge and understanding and if I lack that then my opinion, however firmly held, cannot be accorded much weight. Knowledge and understanding provide the foundation; without them opinion is rendered flimsy, worthy of polite note rather than serious consideration.

Wine-matching advice of every stripe flowed in torrents when the tasting menu came on the scene a few years ago. Initially it was an exciting innovation, an opportunity for chefs to showcase their talents courtesy of a merry-go-round of mini courses that delivered hits of gustatory dopamine via ecstasy-inducing flavors. Each was a shooting star, gone before it was properly seen, to be followed by another incandescent morsel. Its time has passed, but it's going down fighting, its allies being the cohort of waiting staff who indulge in the "raised pinkie" service ritual, where the waiter uses a little finger to introduce every morsel, much like an art expert might explain a painting. Where a thrusting forefinger looks accusatory the pinkie has a primness in tune with the dots and drizzles of each course. The digit sweeps and points to focus the verbose introduction. A users' manual would serve as well and could be ignored as necessary. As the waiter wades through the learned-off-pat explanations good manners suggest a modicum of attention, but fifth or sixth time around my patience frays.

Tattoos

I am baffled by the recent worldwide popularity of tattoos—whose charms chefs are notably susceptible to. Perhaps this is to signal an untamed spirit that challenges norms and refuses to toe a preordained culinary line. If so, yawn. At least chefs are in the kitchen, their front-of-house colleagues are interacting with the customers, putting the frighteners on them as they reach to serve and a cuff rises to reveal a serpent crawling out of hiding. Tattooed sommeliers enjoy even greater opportunities for unsettling diners as they do the round of the table pouring for each guest. Tattoos with dinner? Not for me.

The famed Chicago restaurant Charlie Trotter's was the zenith—nadir?—of this school of waiting. Saint Patrick's Day in the Windy City a few years ago called for a proper celebration, so Trotter's was chosen. The gloomy décor did not presage a headline evening, and a couple of hours later, wearied by effusive explanations and gustatory dazzle, I contemplated a snarl to banish the waiter but settled for a shoulder sag and head droop. He didn't break drone; the building could collapse, but he would finish his soliloquy.

Sommeliers also benefited from the tasting menu, as it encouraged them to suggest outlandish wine choices. We may not mourn the passing of the haughty sommeliers of old, but some of his successors need a wing trim. Good sommeliers listen and observe; they are not upselling con artists. They might massage a wine choice in a favored direction, but any further is charlatan territory. Too many of today's sommeliers are knowledge-burdened obsessives, high priests of food and wine matching who spew facts and figures like talking clocks, information without illumination. Some announce themselves with a straight face as "director of the wine program" and state they have "curated the wine list," which makes them every bit as pretentious as their haughty predecessors. This is self-importance gone mad and reeks of pompous puffery. It's a wine list, not an art gallery stuffed with old masters. Wines are selected or chosen or picked, not curated. The tasting menu encouraged them to run amok, to showcase vast knowledge, fact piled on fact. Too often they favored quirky over mainstream, eccentric over classic.

Mercifully, not all today's sommeliers are afflicted with information overload. The best are a joy to interact with, adding to one's enjoyment and understanding. Meeting them is like a breath of cool air in a maelstrom of babble: assured, confident, and not possessed by a craven need to impress. For people who care about

Great Grape: Grüner Veltliner

Buying a bottle of Austria's signature grape was once the vinous equivalent of playing Russian roulette. It might be bland and insipid, it might be jagged and toe-curling, it might be delicious. Weirdly, it might even be all three at once, a jangling amalgam of opposites, vying for supremacy on your palate. No grape could have a domestic squabble in public like Grüner. Not many gave it a second try.

It is now time to do so. Grüner's day has come and it currently enjoys cult status among wine aficionados, while only a fading echo remains of the querulous individual upon whom grace had to be forced. Grüner has character—in the form of a spicy kick at its core, akin to the prickle of rocket in a green salad. When counterpoised by pure fruit flavors this makes it a great food wine, responding, chameleon-like, to the flavors in a dish. And nobody should doubt that Grüner ages well. The best can keep thirty years, deepening into canary yellow splendor with abundant succulence and mild nutty notes on the palate. It was once joked that winemakers used to make it taste as awful as possible—and then wonder why nobody liked it. No more.

making a good wine choice in a restaurant, it is hard to overstate the comfort of knowing you are in their hands.

Where the tasting menu gave less gifted sommeliers too much latitude, their more competent peers recognized the challenge and rose to it with considered wine pairings, shimmying impressively between the myriad challenges. However, even they must have balked at the arrival of sharing plates—pity the sommelier tasked with suggesting wine matches for a slew of them. They may be all the rage, but I am passing on this trend. I am not sure which is worse, the right-on vibe and the attendant peer pressure to buy into it, or the passing of dishes this way and that with the resultant mishmash of flavors and inevitable mess on the table. I prefer not to flip-flop from one dish to another, playing culinary speed dating. It is fine for a grand alfresco lunch where bonhomie and celebration take center stage. That's its place, so leave it there.

The sterility found at the other end of the dining spectrum—where every morsel has its place in a regimented parade—prim segments of food, cut and crafted, dressed and drizzled, presents a tableau that simply *must* be photographed. Irritation increases when the most ardent snapper at the table insists on photographing every plate, so everyone waits their turn, tilting the plate for better angle and exposure. The prettified food is adored more for pixel count than palate impact. All for a transitory moment of attention on social media.

Choreographed service goes hand in hand with prettified food, fine so long as it does not intrude and become a floor show. If it is regimented and inflexible it defeats its purpose—to be seamless and invisible—and becomes a straitjacket into which the diners must fit. The service should fit around them like a comfort blanket, not hem them in like an electric fence. If it is over-drilled it detracts from enjoyment, distracting attention and interrupting conversation.

Diners need to play their part too. I have great sympathy for chefs forced to execute backward somersaults to accommodate the whims and wishes of fusspot diners who suffer from every manner

The Allergen Index

How did we survive without the Allergen Index? The ever-lengthening lists of naughty-but-nice components in every dish, listed at the bottom of an increasing number of menus. There's an ingredient to upset every taste, each assigned a number or letter, clusters of which are listed after each dish, looking like clues to a crossword. A simple rule of thumb for maximum enjoyment is to make your choices based on the number of offenders, the more the merrier. To add further confusion the calorie crunchers are insisting their figures be thrown into this muddle. Menus are becoming a code-breaker's delight but they are charmless and utilitarian—dining by data.

of allergy and affliction, real or imagined. They have their work cut out, trying to deal with oddball requests too often voiced as puerile, foot-stamping demands.

With greater justification, today's diners should direct their attention to the myriad design catastrophes that pose as cutlery and crockery. Is it too much to expect utensils designed primarily with purpose in mind? Can function not be given precedence over form? Knives with circular handles that rotate in your grip if any lateral pressure is applied to the blade are a minor irritation; those that fall blade up when rested on a plate are a danger. Who thought they were a good idea?

The biggest sinner is the bowl-plate, a broad basin into which your cutlery slides if not placed against a solid item of food to arrest the slide into gooey-fingered retrieval. Perhaps worse is when other items, never intended as food receptacles, are repurposed for service on the dining table:

From the house roof, the ton-weight slate.

From the supermarket, the shopping trolley in miniature.

From the dairy, the galvanized bucket, also in miniature.

From the larder, the jam jar with fiddly lid.

From the quarry, the stone slab, requiring a weightlifter's grip to carry.

From the sawmill, the wooden slab, crafted for heft, not grace.

And from the doctor's surgery comes the syringe, sans medication but filled instead with some potion to be injected when instructed.

This "slates and slabs" pantomime turns dining into farce. I would say to restaurateurs: If it's quirky and seems like a really clever idea, don't use it. If it fires your imagination with how it might be utilized to present your food, don't use it. If you think it will make a memorable statement, don't use it. Remember the staff, they have to wrestle the hardware, and remember the diners, they're paying your bills. Try using the fancy stuff yourself first. Try retrieving your knife from a puddle in the giant bowl-plate, wiping off the handle and starting again. See how much fun it is to spend the remainder of the meal fearing a recurrence. This is dining on eggshells. The plain old circular white plate may be a triumph of function and form. What a triumph.

Dairy-free, wheat-free, gluten-free, nut-free, fat-free, salt-free, sugar-free . . . joy-free. On and on the puritan parade goes. This is counterfeit food, where the essential ingredient—that which gives a foodstuff its identity and appeal—is removed to yield a sham version of the real thing. Today, excusing those with genuine allergies or ailments, too many people insist on this hobbling of food. If all the print was removed from a book, to leave a sheaf of blank pages, would it still be called a book? Yet we are happy to remove fat from cream and caffeine from coffee, the elements that give them their creaminess, their coffee-ness, and still label them as cream and coffee.

Removing essentials from foodstuffs is one thing, but removing "black list" components from actual dishes is a higher level of crime. Step forward the egg white omelette, the roofless house of the culinary world. It's a complete aberration. The customer may always be right, but a diner asking a chef to commit such a criminal act should be made to wash the dishes. We subject ourselves to these gustatory travesties in a deluded search for a healthier lifestyle. It would be simpler to limit our intake, to indulge occasionally.

In an effort to mimic this madness, some wines have decided to drop the alcohol. In a world where all the other aberrations are accepted as valid choices, no eyebrows raised, it is not surprising that de-alcoholized wine is allowed along for the ride. It's the wine that dare not speak its name, the engine-less motorcar of the drinks world. Would we buy an engine-less car? It would look like the real thing; you could sit in it and enjoy the view, fiddle with the controls, turn the wheel, play at driving, wave to your friends and go "vroom, vroom" as a child might.

There is a school of thought that accords it some validity, arguing it should be invited to the party, but it would feel more at home in a puritan church gathering. Let's not be vague: de-alcoholized wine is an imposter, a blight on all that is good about wine. It is a neutered beverage dressed as wine. Have a glass in the evening and wait for it to unravel the day's cares, and you will wait in vain. If I don't want to drink wine I'll drink water. I am a huge advocate of lighter wine styles but not those that are bashed about to remove the alcohol.

Diners who insist on neutered wine and counterfeit food are a nightmare for chefs, who are expected to pander to these fusspots, a legion of whom has been created and encouraged by ever more forensic analysis of food and drink. Restaurant dining should be fun, a guilt-free shared pleasure, not a dismal exercise in ingredient-checking and calorie-counting. Those strictures belong at home, if they belong anywhere; chefs should not be handcuffed with them. That's not what we are paying them for. They should be allowed to cook untrammeled by fads of the moment, beliefs that will be ditched as soon as the next fad rounds the corner. Why force them into spending time on a pointless exercise, like putting calories on a menu, that has nothing to do with cooking? Why are they tasked with a responsibility that should be ours? Diners must look after their own waistlines, not expect others to do so.

Crubeens and Stout

Time was when this was a combination for sleeves-rolled-up guzzling, with dribbles of fat and meat juice running down the chin. This is how I first enjoyed crubeens—as pigs' feet are called in Ireland—while in college. One of the lecturers would boil a cauldron of them for twelve hours on the eve of the annual Colors Match against our rival university. Then they were heaved into an assortment of bowls, and you helped yourselves to a gelatinous lump, trying to avoid scorching fingers and scalding tongue as you searched for the few nuggets of flesh among the fat and bone. Pints of cool stout salved burns, the bitter tang sluicing away the fat, leaving us in fine fettle for bellowing support of the team later.

Since then, crubeens have been to finishing school, shorn of rustic coarseness, before presentation in the genteel surroundings of linen-and-crystal restaurants. The raucous delight is gone, replaced by prim, measured enjoyment. Thus gentrified, trotters are crafted into decorous shapes, politely packaged nibbles. They are easier to handle, may even taste good, but sterile consumption is not their destiny. They are for gluttonous feasting, not prissy dining.

Another demand du jour insists that menus contain some sort of mission statement about sourcing and sustainability, which is a sham unless you apply the same strictures to yourself day in day out. You can't don your I-care-for-the-planet-and-animal-welfare hat only when dining in a restaurant. Indeed, it would be better to wear it at all other times and then permit yourself some indulgence when you dine out. Allow restaurant dining to be an enjoyable treat and not a guilt-racked search for the least offending, most ethically superior item on the menu.

* * *

In all the heat and dust generated by evanescent fads and foibles, one area of food and wine matching fails to get the attention it deserves—difficult pairings. This is where guidance is badly needed, yet there is a dearth of advice on how to deal with foods that don't automatically suggest a wine match. These "guerrilla" foods are liable to snarl if not accorded due respect by way of a considered wine match, one that won't be cowed by their behavior. Top of the pile are andouillettes, those intestinal sausages that reek of their origins and which are only for diners sound in wind and limb and not worried about the anatomical source of the dish. All others should look away and order the green salad. Andouillettes are grown-up food, fumes made flesh. As such, they require a wine with some "sting," something to lance the fumes and stop them engulfing one's being with pungent aromas and feral flavors, a wine that can bite back if necessary. Step into the ring crisp young Chablis, snappy Chenin Blanc, or piercing Aligoté, all of which can put manners on them. The trick is to get food and wine dancing together, not wrestling. Get this match wrong and the andouillettes will trample the wine to extinction. Kippers are similarly challenging, oily and reeking of smoke. An *en rama* Fino Sherry, robust and unfiltered, might do the trick, though for stouthearted types only a rugged Islay malt whisky will do, even at breakfast time.

Cheese and red wine is not the match made in heaven that many people suppose. They tend to get rammed together after a meat main course when there is some red wine remaining, but this is lazy and often unsatisfactory. Pungent cheeses such as Époisses, and blue cheeses led by strident Roquefort, slaughter red wine. The latter's salty hum is best when cosseted by Sauternes. There is no such obvious match for Époisses, perhaps a buttery rich Meursault, though my best effort was an aged, top-notch Alsatian Pinot Gris. My Burgundian friends react in horror when I tell them I start the day with Époisses, spread thick on a baguette for breakfast and accompanied by strong black coffee. Red wines such as claret work best with hard cheeses. Comté is a popular choice, and there

Perfectly Simple

Blood sausage or black pudding may not be to all tastes but when it is good, properly moist and not crumbly dry, it is a treat, humble not sanctified. In Ireland it is normally confined to breakfast duty, but on a visit to an Argentinean winery a few years back it formed the centerpiece of lunch. A cannonball of pudding, black as coal, was rolled around the barbecue until crusty and cracked. Heaved onto a carving dish it was sliced open to reveal a gooey center that slid across the plate, to be mopped up by a hunk of bread and washed down with a no-holds-barred Malbec. The wine was young and as vivid a purple as the pudding was black. Greed conquered decorum as seconds were requested. Context is everything.

are many others worth considering: Gouda, Cheddar, Beaufort, Mimolette, Coolea. Nothing too assertive or overripe.

The "matches made in heaven" that send foodies and wine lovers into raptures comprise a celebrated roster: beef and Burgundy, foie gras and Sauternes, lamb and claret, Stilton and Port, lobster and Meursault. Beef is probably the most forgiving of these, for it can be enhanced by a broad roster of red wines and not just Burgundy. It is equally comfortable with Rioja, Barolo, numerous great Cabernet Sauvignons from around the globe, including claret, and great northern Rhône Syrah. It presents a broad target for gustatory matchmakers, unlike lobster where the focus must be narrower, and usually Chardonnay-based. It's not a tricky food but there is a smaller sweet spot to be aimed at; the fit must be neater if the match is to sing.

A regularly encountered food and wine matching challenge is the western world's traditional Christmas dinner, which is a sore trial for wine, a gustatory assault course with hurdles all the way. The roll call of its constituent parts is daunting: turkey and ham, sage and onion stuffing, sausage stuffing, potatoes done two or three ways, sprouts and other vegetables, bread sauce, cranberry sauce, gravy. Most wines quake when faced with such largesse on one plate. What is needed is a wine that can ride with the punches, deal with the pressure yet still remain in good fettle, a wine with an ample flavor and some "cushioning" richness. The choice is as much about texture, which should be smooth, as flavor, which should not be sharp. In red wine that means fruit and not too much tannin, and in white—which should not be ignored as a match for the Yuletide meal—fruit and not too much acidity. Amplitude in both goes without saying. Step forward Châteauneuf-du-Pape, the famed wine of the southern Rhône. In both its red and white versions it is lavishly flavored and easily up to the challenge. The perfect solution is to serve a glass of each side by side, from a famed estate such as Clos des Papes, and let your guests decide on the best match.

In Australia, where Christmas falls in high summer, with the barbecue going full tilt, a favored choice is sparkling Shiraz, a wine whose frothing exuberance can swat aside this most challenging meal in a cascade of inky bubbles and licorice-sweet fruit and, unless ferociously chilled, can do a similar job on unsuspecting jowls and palate.

Perhaps the most important advice when choosing Christmas wine is: do not, repeat do not, save your most treasured bottle for the big feast, the occasion alone will swamp it, never mind the food. The circumstances are all wrong for appreciation of a special bottle, perhaps a cherished gift from years previously, received to mark a birthday or anniversary. Oil and water have more affinity than special bottles and special occasions. Let the bottle be the occasion, open it on one of the quiet days after Christmas when it won't be swamped by all the hoopla and fuss of the big day. And, whatever style it is, don't be tempted to serve it with elaborate, richly sauced, or highly seasoned foods. Let it take center stage and accompany it with something simple. The best wines need a good accompanist to help them shine—not a rival diva. Keep it plain.

Another groaning plate that could destroy any wine is the breakfast "fry up," but thanks to the early hour at which it is served most wines escape that fate. This gargantuan slamming together of sausage, bacon, and fried egg, with bit players such as baked beans, fried mushrooms, and grilled tomatoes, is the Ugg boot of the culinary world—comfortable, ubiquitous, but hardly stylish. Yet it is placed on a pedestal as a culinary triumph, exempt from critical judgment. Perhaps it escapes censure because of the weight of tradition behind it?

The "fry up" is fodder-as-food and because early morning is when we feed rather than eat we mislay our critical faculties when considering it. I enjoy it, but to consider it a culinary triumph is way wide of the mark. Culinary triumphs involve some degree of creativity. Anyone can do this agglomeration of foodstuffs. If a wine must be served, sparkling is the best choice, perhaps a vigorous young Champagne, not a treasured bottle. The bubbles help smooth over the humps and hollows on the plate.

Champagne is a far better accompaniment to food than is generally supposed. It is almost never served through a meal, yet its innate exuberance makes it versatile and adaptable. Great Champagne deserves its place at the table, yet it struggles to escape from the aperitif-and-celebration pigeonhole into which it is cast— often by its producers who are only now beginning to see the error of their ways. This, the noblest of wines, partners brilliantly with humble fish and chips, as well as with exalted Beluga caviar. For the latter, an aged, top vintage of Dom Pérignon or other member of Champagne's aristocracy, sumptuous and toasty, is what's needed to counterpoise the creamy saline taste and satin texture of the caviar. Eschew all trimmings to allow the splendor of the dark little beads, shining and winking, to show at their best, rivaling the Champagne for palate impact before combining with it to create one of the greatest of all food and wine matches. Unless your fairy godmother is paying, Champagne doesn't need to be paired with caviar. Its versatility may not match Fino's, but fashion is at last swinging in its favor, and it is now often served in proper wineglasses and is at least being referred to as wine. Champagne is coming in from the cold, but much still needs to be done to free it from the aperitif-and-celebration grind.

Steering a sensible path through the food and wine matching minefield is a challenge, yet it can bring ample reward. A proper match clarifies flavors, particularly in the wine. Most wines benefit from being paired with food and some—the European classics in particular—are positively enhanced by a good match. Drunk on their own, they can be lean and slightly aggressive, until the food pulls the flavor into focus, softening tannins and smoothing acidity. It's worth making the effort—and important not to get consumed by it.

11

GOLDEN WONDERS

Dublin Castle, Ireland, Saturday June 15, 2019, one hundred years to the day after John Alcock and Arthur Brown completed the first transatlantic flight, landing in a bog in the west of Ireland. To mark the anniversary I was commissioned by Avolon, one of the world's largest aircraft leasing companies, to organize a celebration dinner for 180 guests from across the globe. They were treated to Dom Pérignon 2009, de Fieuzal *blanc* 2014, Léoville Barton 2000, and Lynch-Bages 2005, all in magnum, yet in biblical fashion the best came last.

The meal finished with a golden flourish, as Château d'Yquem 2009 was poured in ample measure. For some guests it was their first taste of this storied wine, one that many have heard of and few have drunk. Telling the diners that Yquem is made from rotten grapes, sticky and shriveled, brought incredulous looks. Surely something so succulent could not be the child of rot? It can, and without the rot it cannot be made at all.

* * *

Château d'Yquem is the alpha male of the Sauternes region, southeast of Bordeaux. There, if the morning mists of autumn are burnt off by the midday sun, and if this happens day after day, the noble rot or *botrytis cinerea* attacks the grapes, causing them to shrivel. The result is a greatly reduced quantity of concentrated juice, which is then crafted into honeyed Sauternes. It is a high wire process that dallies with failure all the way. So much is out of the winemakers' hands, all they can do is hope for the oscillation from fog to sun. Each day adds its increment, turning the fruit into unappealing grape putty. Yet locked inside is essence, hidden there by the weather gods, and now it is the winemaker's job to pan for gold, to coax magic from mush.

Top Trio

The wine club dinner finished with a flourish, a trio of Château d'Yquem: 1988, 1989, and 1990, probably the greatest vintage trio of Yquem ever. The wines were paired with classic accompaniments: a piece of foie gras, a miniature tarte tatin, and a sliver of Roquefort, which formed an equilateral triangle on a large circular dinner plate, plain white and completely unadorned by drizzles of this or garnishes of that. The guests were asked simply to comment on the wines, not rate them in order of preference, and to see how they matched with the food.

Murmurs of delight dominated the conversation. Each wine was an exemplar, separated only by a whisker from its sibling. Distinguishing features were minute, but gradually consensus emerged that however good the 1988 was, the 1989 was a scintilla better and the 1990 a whisper better again. There was only a hair's breadth between them, but moving from first to last there was an impression of slightly greater volume and a trifle more length. Repeated rechecking confirmed that opinion.

When the châteaux of the region were classified in 1855 Yquem was accorded a rank above all others, as the only *Premier Cru Supérieur*. Its geographical position reflects its standing—the château occupies a hilltop, gazing down on its neighboring châteaux, clustered around like acolytes: Suduiraut, Rieussec, Guiraud, and others. They are famed in their own right but none is so famed as Yquem. How to describe it?

It combines a gallery of supreme flavors: the sweetest peach, the finest heather honey, caramelized orange, melting butterscotch, vanilla custard, and baked cream. Then come the minor players, each adding a few pixels of delight: sweet spices redolent of the Orient; sensual licorice; mango, pineapple, and apricot from the tropics; preserved lemon flashing on the palate; creamy almonds; fudge and toffee. The alcohol spins this cornucopia into gold, the flavors melding to yield a collage enjoyed in one sip, one mouthful of essence. Some would say it's a taste of heaven, a polished pearl of flavor, ringing and reverberating, unforgettable. Château d'Yquem is the greatest Sauternes and, some contend, the greatest sweet wine of them all.

Despite such renown, and the superb quality of other Sauternes châteaux, the wines remain stubbornly out of fashion. Sweet wines are emphatically not having a moment now, shunned by all but a hardy bunch of aficionados. In response, numerous châteaux also make dry wines. Yquem started the trend back in 1959 with "Y" (pronounced "Ygrec"), though it was some years before others followed suit, many named with a nod to Yquem. Thus we have G de Guiraud, S de Suduiraut, and R de Rieussec, while Château Coutet used its own imagination, naming its esteemed dry wine "Opalie."

Perhaps sweet wines are shunned because we are "over-sugared"—at every turn we can indulge in a sweet treat, it is almost impossible to avoid saccharine overload in our diet. A couple of hundred years ago sweetness was a prized luxury. It is much easier to get our sugar hit today—witness the walls of sugar, tricked up as every imaginable confectionary, that assail the eye at countless

Great Grape: Sauvignon Blanc

If grapes used social media, Sauvignon Blanc would have more followers than any other. It is hard to account for this popularity, save to acknowledge that when Sauvignon is good it delivers a pure and vibrant flavor that rings like a bell on the palate. Forty years ago few people had heard of Sauvignon, when along came New Zealand, and things haven't been the same since. Kiwi Sauvignon—racy and riveting—conquered the world, but rampaging international success has not been all good for Sauvignon. Legions of imitators hopped on the bandwagon to churn out an ocean of insipid versions that had all the charm of cold tea. The vibrancy was lost in a sea of washy flavors.

Sauvignon doesn't always sing solo. In Bordeaux it duets with Semillon to yield the world's most under-appreciated dry white wines in the Graves region, while down the road in Sauternes and Barsac it is an essential building block in some of the greatest sweet wines on the planet. Soloist and team player, there is more to Sauvignon Blanc than first appears.

retail checkouts. Whatever the reason, eventually the market decides and if the market is no longer interested in sweet wines you must adapt to stay in business, which makes the strategy of going "dry" understandable. It also has some attraction when the fraught nature of sweet wine production is considered. The winemakers can only monitor the weather and implore the heavens. But eschewing sweetness is like an opera singer deciding to switch to speaking-only roles. The singing voice is what gives them identity, what makes them special. Without it they slip back into the crowd, their essence abandoned.

However celebrated Sauternes may be, Tokaji from Hungary has a more impressive ancestry. Many wines vie for the title of noblest, most aristocratic of them all yet none can stake a claim as compelling as Tokaji. The fabled sweet wine is produced in the foothills of the Zemplén Mountains in northeast Hungary and stories are legion about how it was coveted by royalty, to the point where the Russian "Greats," Peter and Catherine, used a detachment of Cossacks to safeguard the wine's journey to their cellars.

About two hundred years before anyone heard of Sauternes, Tokaji was being made from nobly rotten grapes and it was the first wine to be made so. Its apogee is Eszencia, made from the juice that dribbles from the botrytized or *aszú* grapes, pressed only under their own weight. Twenty-five kilos yield about one liter of juice, which oozes from the murky mess, viscous drop by viscous drop. Thereafter the liquid toys with fermentation to yield a faintly alcoholic and abundantly, ludicrously sweet wine. This can take years and the result is syrupy and dense and virtually ageless, with a life span measured in centuries. Thanks to its rarity it has been credited with miraculous powers, though only monarchs and plutocrats have been in a position to test such claims. It is clichéd to describe Eszencia as "nectar," yet that's what it is.

At a less-exalted level, Tokaji Aszú is the wine most likely to be encountered by ordinary mortals. It is still triumphantly sweet but without the massive density of Eszencia. It is made by blending *aszú* grapes, from which the Eszencia has dribbled, with must or fermented wine made from grapes unaffected by botrytis. The sweetness depends on how many *puttonyos*, or twenty-five-kilo hods of *aszú*, are added per barrel of wine. Five or six "putts" will satisfy even the sweetest tooths.

Tokaji fell from favor after World War II, as the drab hand of Communism fell across eastern Europe. The Communists were not kind to the regal wine and for over four decades quantity was prized above quality to produce a soulless version, wine as elevator Muzak. It was stripped of character; adding a few spoons of sugar to a dry table wine would hardly have been more dismal. Great vinous art was turned into a painting-by-numbers exercise.

In time, that period will be seen as another chapter in a long story, and thankfully the most recent chapter is as good as that was bad. Since the fall of Communism, outside investment and the return to private ownership of many vineyards has led to a rebound in quality. The painstaking practices of old have been reinstated, and the results are thrilling and deserving of far greater notice. Revival is in the air, fresh life is being breathed into this classic with every passing year, and mere words are challenged again to capture its beauty.

Tokaji shimmers like amber with glints of pale copper, and gives off scents of apricots and peaches, with a firmer note of caramelized orange and spice. The flavor unfolds in a rolling wave, luscious and honeyed but not cloying thanks to a tangy zip of acid. It plays with fire, threatening to tumble from caramelized to burnt, but never crosses that line. Freshness and richness form the warp and weft in a satin-textured wine that doesn't have the same licorice quality as Sauternes; there is more "spark" on the palate. That's Tokaji.

It is time to speak of Riesling. Where Sauternes has pillowy flesh and glides to a glorious, trumpets-sounding finish—and where Tokaji sings with higher top notes and not as large a chorus—Riesling is different again. Its profile is slimmer, an electric intensity couched in succulence, which causes it to dance rather than glide across the palate. Sauternes calls to mind full flavor triggers of honey and custard; Tokaji is less caressing, marked by incense-like spices; Riesling, in its sweet incarnations, majors on concentrated fruit essences almost shocking in their intensity. Without the grape's acidity they would overwhelm—with it the flavor is close woven, held taut, resulting in wines more poised and less expansive than the other two.

Sweet Riesling is a transnational treasure, and commendable examples are made across the globe. Many are singular expressions of fruity sweetness, delicious in a simple way, melody without harmony. Haunting greatness is more elusive and tends to be the preserve of long-established regions, such as France's Alsace and Germany's Mosel. In each, Riesling answers a different call but, such is its distinctive voice, no amount of geographical variation or winemaking wizardry can silence it. Geography can shape it and winemaker nuance it, but it is still Riesling. Emphatically so.

Alsace feels more European than French and distinguishes itself from other French wine regions by using the name of the grape to identify the wine. Riesling is its most planted variety, by a whisker from Pinot Blanc, with Gewurztraminer not far behind. Some contend this latter grape is Alsace's most noble, letting narrow-minded chauvinism trump clear-headed reason. Gewurztraminer is Alsace's signature grape, the region's speciality, but Riesling makes its greatest wines. No question.

Styles range from bone-dry to lusciously sweet, and Alsace is esteemed for them all, but it is the latter that speak of their origin with the strongest accent. In short, it would be easier for a winemaker elsewhere in the world to imitate a dry Alsace Riesling than a sweet one. They come in two categories: Vendange Tardive and Sélections de Grains Nobles, the first made from ultra-ripe grapes and the second from even riper and usually botrytized grapes.

In the best, Riesling's delicacy is retained while the intensity is ratcheted up by the cloak of sweetness. Flesh and substance come from ripeness and not oak, certainly not from the big oval casks, seasoned veterans of dozens of vintages, that are a feature of the region. Riesling and oak don't step out together, neither in Alsace nor elsewhere. It needs no props.

One of the greatest I ever tasted was Trimbach's Clos Sainte Hune, Vendange Tardive 1989. Sainte Hune is one of the world's greatest wines, resonating with inestimable grace—in its dry version. That resonance rose to new heights in this wine of fathomless depth. It shone ocher-yellow in the glass and a swirl released an exotic aroma, leading onto a full palate, replete with marmalade and candied fruits. It was complete and harmonious, with memorable purity and is now a wine of legend, esteemed the world over, the best flag-carrier Alsace could want.

* * *

As Riesling moves northwest across the border into Germany's Mosel it sheds weight and gains finesse. Where Alsace has ballast, Mosel has filigree flavors, deceptively transparent, yet not frail. Delicacy should not be mistaken for timidity. They have supreme lightness of touch allied to hidden vigor, and the best leave you wondering if you ever want to drink anything else—yet their fate is perplexing.

It was not helped by the heavy-hoofed bureaucrats who, half a century ago, concocted the shot-in-the-foot law that allowed all sorts of sugar water masquerading as wine to be foisted on a sadly too-willing public. That blunderbuss legislation destroyed Germany's reputation, yet in the hands of conscientious winemakers the great sweet wines retained their quality. They couldn't, however, escape the tarnish that sullied every German wine as the twentieth century

Riesling's King

The visit started slowly. It was as if Egon Müller, immaculately dressed and precision mannered, was assessing his nine visitors before opening any treasured wines. One bottle was proffered and tasted. Then another. A spittoon was asked for and reluctantly provided. Then another bottle. Every time Müller went to fetch the next bottle he was gone for a little longer than previously— digging deeper into the cellar we supposed. We were right.

The wines were linear and precise, crafted by a jeweler's hand, each flavor facet a delight in itself and indescribably beautiful when allied to the others. The increasing sweetness levels added depth and luster as we moved up the scale until we were rejoicing in a Trockenbeerenauslese 1975, the requested spittoon long forgotten. It unfolded like a Russian doll, a glorious cascade of flavor, layer upon layer of gemlike succulence that echoed and echoed on a finish that seemed infinite.

closed, with the new world triumphantly recruiting legions of young drinkers who cared nothing for tradition or the backbreaking toil that lay behind these wonders.

Getting to grips with German wines starts with the labels, whose bark is worse than their bite. They are wonders of logic if not simplicity; patience is needed to decode them and to unravel the gothic script and heraldic flourishes. Until then, they remain brain-busting. Thereafter they only snap occasionally, such as when the "trocken" trap is encountered. It means "dry," yet the sweetest German wines are labeled Trockenbeerenauslese—in this case the trocken refers to the shriveled grapes from which the wine is made. Confusingly, a Spätlese Trocken is dry because, here, "trocken" refers to the style of the wine.

This can be a trap, for a Spätlese wine will usually be slightly sweet and "trocken" is only added to the label if it is fermented to full dryness. A turn-of-the-century fashion swing to dryness saw German winemakers eschewing the traditional leavening of residual sugar in favor of ultra-dry wines. These had something of a moral-high-ground attraction but they verged on shrill unless impeccably made. Thus, trocken now wears two hats; where it is placed on the label is the clue.

Classic Mosel Riesling is divided into six categories: Kabinett, Spätlese, Auslese, Beerenauslese, Eiswein, and Trockenbeerenauslese, in ascending order of sugar ripeness in the grapes. Auslese is the transition style: top of the first three as well as bottom rung on the ladder to uber-sweetness. Above it comes a luscious magic kingdom topped by Eiswein and Trockenbeerenauslese. The difference between these is that the sugar concentration in Eiswein is achieved by allowing the grapes to freeze on the vine, and in Trockenbeerenauslese by botrytis. And, however good Eiswein may be, the botrytis weaves magic into Trockenbeerenauslese, a hedonistic dimension that elevates it to an unrivaled position among all sweet wines. To wring such wonders from the Mosel's vertiginous vineyards is a winemaking feat like no other.

Simply working the slopes and tending the vines is a grueling task, and only when conditions allow the grapes to achieve super ripeness, enhanced by botrytis, can the most special wines be made. Rugged and harsh origins in the vineyard somehow yield delight in the glass. From one to the other is the greatest span of achievement in the winemaking world.

A great Trockenbeerenauslese, or TBA, is wondrous, it spirals on the palate, revealing facet after facet of flavor. Minutes pass as it lingers. Reflective silence is the best tribute to pay when such a wine is encountered. Its hallmark is thrilling intensity with no aggression; everything revolves around the acidity, it is the center of gravity, like a vanishing point in a painting. TBA is genius wine: the most poetic, the most musical, the most artistic in the world. Nothing rivals it for unfathomable excellence and sensual delight. Thrill and tingle, vigor and insistent energy are its calling cards, rather than the luscious tongue massage delivered by Sauternes. Each to his own. I lean toward the Riesling.

Europe does not have a monopoly on wines crafted by painstaking effort to yield minuscule quantities of palate joy. Wherever wine is made there are nutcase winemakers willing to enslave themselves in pursuit of lush excellence. From a packed field two stand supreme, one South African, one Australian.

Rutherglen Muscat is uniquely Australian; it isn't modeled on an old world template. It's a national treasure, though it is hardly treated as such—Aussies may make in-your-face claims about the delights of Barossa Shiraz or Yarra Pinot but are silent when it comes to championing the joys of this true original. Rutherglen sits in the northeast corner of Victoria, a couple of hours' drive from Melbourne, with speed cops monitoring your progress all the way. There, the sophistication of Australia's most appealing city is left far behind but any sense of loss is rectified by a luscious sip drawn straight from a giant cask.

The contrast between these ancient vessels and their contents is huge. They are scarred by time, cobwebbed and cracked, with treacly dribbles seeping from the joints. Chalk marks identify the contents and guide the cellar master as he climbs atop the casks to draw samples via a slim steel cup on a string. Dropped through the bunghole it sinks and fills, to extract a measure of glowing walnut liquid, burnished by flashes of amber.

Treacle and preserved fruits—dates, figs, prunes—glide across the palate, leaving a long, lavish echo. Amazement follows when the production process is explained for, as with Madeira, it calls to mind the tempering of steel rather than winemaking. It begins and ends with heat: sunshine bakes the grapes on the vine, turning them almost to raisins, and when the wines are made they are transferred to the ancient casks to endure torrid years under tin roofs that afford scant protection from the heat.

It is difficult to believe that grapes and wine, treated so, could yield such a precious essence. After this trial by fire comes trial by time, which continues almost indefinitely and leaves the wines reinforced, not diminished. As years slide into decades they gain depth and richness, and immortality, at least when measured against a human life span.

Immortality of another kind is found some 6,000 miles west of Rutherglen in Constantia, the bucolic suburb of Cape Town where South Africa's first wine was made in 1659. A century later "Constantia" wine, sweet and luscious, was vying with Tokaji as royalty's favorite, and later still it eased Napoleon's torpor as he whiled away his exile on Saint Helena. Yet by the twentieth century it was gone, as extinct as the dodo. All that remained was the renown, it was a treasure lost to history, as was its regal customer base—the emperors and kings—but while few yearned for their return some dreamed of bringing the legendary elixir back from the dead.

Today, it has been resurrected by Klein Constantia (say Klane) as Vin de Constance, a faithful revival of the eighteenth-century classic. In the nineteenth, the original succumbed to phylloxera and, with many of its customers succumbing to more violent fates, it slipped away without fanfare. Today's reincarnation seeks to mimic the original in every respect—even down to the bottle shape, a reproduction of a survivor from the glory days. Long necked and slightly dumpy, with a distinctive "wobble" in the base, it looks like it is starting to melt having been left on a hot surface.

Launched in 1986, Vin de Constance quickly won a place in the sweet wine firmament and is now South Africa's glorious contribution to that niche world. Its style can be partly described by what it is not: it is not syrupy or honey-fleshed, nor is it spicy, nor is the fruit burnished by botrytis. The flavor register is lighter yet still penetrating, orange blossom rather than caramelized orange. Such finesse gives it immediate appeal, particularly for palates new to sweet wines, and it disappears out of the glass more rapidly than its lavishly flavored cousins.

Just as a coat in winter fends off the cold, so the sugar in these wines resists time's corrosive attention, preserving them for generations. They are held in a saccharine embrace, supported as the years pass, the flavors allowed to develop and deepen, the texture to gain luster, and the whole to achieve a rich patina, all without spoiling. The secret, however, is not the sweetness but its opposite, acidity. Just as light is counterpointed by shade so is sweet by acid. A sweet wine without a balancing pull of acidity is a gross nectar that lambasts the palate before scurrying to anonymity. Such acidity-free wines are vinous marshmallows, wobbly puffs of flavor that might best be frozen into ice pops.

Sweet wines, though stubbornly unpopular, have had miraculous powers attributed to them. Tokaji leads the field, capable, according to the more outlandish claims, of reviving the dying. Legend sees it raising monarchs from their deathbeds, to vanquish enemies in battle and father a gaggle of children before finally succumbing to the grim reaper. Thanks to a drop of Eszencia they were given one last gallop at life. Questioning such claims, hairsplitting them to extinction, is a dull pursuit. Far better to admire the imagination that conceived them. In pursuit of everlasting youth perhaps we should drink more of these jewels?

When we do let's do so properly by serving them in decent glasses. Even if only a small measure is being enjoyed that's no reason to use a miniature glass; it's not a crime to use a regular glass. The true crime, the vinous capital offense, is to serve these beauties in thimbles, such as the pinched little schooners used for years in genteel households to serve Sherry, corsets as glassware, and as unforgiving on the contents as the real thing was on ladies' waists. Throw them out.

12

WHAT NEXT?

Crystal ball–gazing never loses its allure. We remain in thrall to guessing the future despite centuries of failure leavened by sporadic success. We persist in looking into tomorrow's murk, striving to get ahead of the game, seeking advantage without effort. The wine world is no different, its players constantly trying to look around the next corner; it is always in flux, sometimes in turmoil, its components shifting and realigning, rendering fortune-tellers' efforts futile. COVID-19 rewrote the future but in unreadable form; flux and turmoil became the norm.

My own track record in wine predictions runs to script, the most wildly inaccurate being the annual foretelling of the "Riesling Revival." This wishful thinking masquerading as conviction has been pedaled as a racing certainty for more than twenty years, yet has always fallen at the first fence. I can offer no explanation for the continued shunning of Riesling in favor of less-distinctive styles. Perhaps people don't want to engage with their wine, content to let it slide across their palates, carrying a hit of alcohol in an unobtrusive package. If so, the future looks bland.

For decades a never-had-it-so-good narrative has dominated the wine conversation, peddled a little too enthusiastically at times. In large part it is a valid assertion, yet unquestioning acceptance of it dulls the critical faculties and opens the door to the bland and the boring, wines so unobtrusive they leave no fingerprints, nothing to comment on. Faultless but vacuous wines have replaced the snarling, palate curdlers of yore, whose demise nobody mourns. These word-perfect wines are akin to a studio recording compared to a live one: manicured, every glitch excised to fit a preordained template. They are "flavor loop" wines, each sip a simple repeat of the previous one, as engaging as nursery rhymes. Wine's variety and diversity is being trampled by the drive to homogeneity.

It has never been easier to make boring wine and, in turn, to make wine boring. This latter crime is committed when wine is presented to consumers in too academic a fashion. In recent years wine education has been championed like never before. Droves of enthusiasts enroll in courses and join clubs, seeking to unravel wine's mysteries. For many it is a truly worthwhile exercise but once it becomes too knowledge-based it diverts into a deceptively attractive path that leads nowhere. If the accumulation of knowledge doesn't endow insight and understanding, it is as futile as piling bricks on top of one another in the hope of building a house. How we approach the study of wine is critical.

The wine film *Somm* told a joyless tale, as it followed four candidates preparing for the notoriously difficult Master Sommelier qualification. They force-fed on knowledge, cramming in fact after fact, to be spewed on cue. It was wine as delight-free fact bomb, every drop of fun wrung from one of nature's great gifts. Super difficult wine education such as this, as well as the similarly challenging Master of Wine qualification, make wine exclusive and give it a cultish aura. (Notwithstanding this quibble, some of the wine world's finest communicators hold these prestigious qualifications.) Yet they are ceaselessly championed as summits of achievement, their intimidatingly high failure rates held up as marks of worth beyond words. Might constant highlighting of the failure rates boomerang back if people begin to ask whether the failure is the institution's and not the candidates'?

This sends the wrong message—that wine is a forbidding topic—and does nothing to broaden its appeal. It also encourages the rise of the wine nerd who feasts on facts; a collection of facts is not an education. Wine should be enjoyed, celebrated, delighted in, not put on a pedestal and sanctified. It's an agricultural product. We should not be too precious about it or invest it with too much gravitas or ceremony; as already stated it must be destroyed to be enjoyed. Above all, the language of wine should not be intimidating, should

Great Grape: Aligoté

"Great grape? Great grape?" The doubters harrumph as they prepare their objections. I once sneered at Aligoté and was only cured by drinking the wines of top names: Pierre Morey, Bernard Moreau, Marc Morey, Laurent Ponsot, Jean-Claude Ramonet, and Domaine de Villaine.

Aligoté is the once-troubled adolescent of the grape world that has now grown up, after a mewling youth. It was tamed in a fashion by Canon Kir of Dijon, who added a few drops of crème de cassis—but that was like giving sweets to an unruly child. Today Aligoté is better cared for in the vineyard, better handled in the winery, and, allied to better ripeness, has lost its rebellious snarl and emerged tangy rather than brackish. In Burgundy, conscientious winemakers take great pride in their Aligoté and make it with the sort of care you imagine they reserve for their grander wines. Silence the sneer and try a glass. The best is yet to come.

not sound like an exclusive argot that only the high priests speak fluently. To achieve this in the future we should approach it with enquiring rather than forensic minds.

The overanalytical approach begets wine trade professionals and educators, writers and communicators whose thinking is constrained by knowledge, where appreciation trumps enjoyment. In a professional setting that is understandable, but for consumers enjoyment is paramount; they are interacting with wine in their downtime, they don't want a lecture, yet too often that is what they get in all but name. Many fact-burdened experts are capable of little more than passing on the burden.

And these same experts are seldom shy about taking a swipe at the wine trade of yesteryear, suggesting that back then you had to have been to the right school, wear a pinstripe suit, and speak with a plummy accent to gain access to wine's golden circle. This tired brickbat is still hurled by the jolly chaps' successors, who should ponder whether one elitism has replaced another. There are more similarities than differences between today's tribe of wine professionals and their predecessors. Pinstriped and plummy may be out, but it has been replaced by a learning-based circling of the wagons.

Wine writing dipped into the knowledge swamp for a time and emerged spewing tasting notes like confetti. The taster-writer adopted an arm's-length, faux-objective stance before penning a tinder-dry note that delivered a quickfire judgment, often based on no more than a minute's assessment of the wine. Has this writing style got a future? I'll stick my neck out and say that its days are drawing to a close. That such an arid prose was foisted on wine lovers and accepted for so long is incomprehensible. Reading about wine should not be a trial. It is a subjective pleasure and must be engaged with to yield up its joys. Being informed that your correspondent's palate has trudged through hundreds of samples on your behalf is groan-inducing. There has to be more to wine writing than wine-of-the-week.

Great Grape: Godello

Nobody knows when Godello will emerge from Albariño's shadow. Its situation is not unique: New Zealand Riesling and Chardonnay, obliterated from view by Sauvignon Blanc, could empathize. They must soldier on, confident of eventual acceptance. Godello, however, does not have the name recognition of the other two, there's no established reputation elsewhere it can hang its hat on, so it must be sustained by the hardy band of followers already cognizant of its charms.

Those charms include a mineral bite and crisp intensity, allied to a fine ability to age that set it ahead of competition from the likes of Verdejo and Viura. Some sing the praises of the former, claiming it is also waiting in the wings, and others the delights of the latter, championing its easy charms. For me Godello is the best. And my money says it will soon gain its place center stage. Get ahead of the curve, and try Godello now.

There are encouraging signs that peak "tasting note tyranny" has passed, the most welcome straw in the wind being the republishing in 2020 of *The Story of Wine* by Hugh Johnson. Three decades after it was written, this magisterial work is more relevant than ever, providing a rich backdrop to the reader's engagement with wine. It comes from the Academie du Vin Library, a publisher dedicated to breathing new life into classic works and also to commissioning new books that "provide the colour, the background and the essential understanding on which any true enjoyment of fine wine depends."

Notwithstanding such developments, wine writers may soon need to retire their pens, or at least expand their skills to include a more immediate form of communication with consumers: online.

In the spring of 2020 the wine world stopped spinning as COVID-19 brought all but essential activity to a halt. Every aspect of the wine business was affected, from growing the grapes to drinking the wine, including the business of keeping consumers informed. Online wine presentations and tastings came of age. Zoom, a name hitherto almost unknown, became the star turn, along with others such as Microsoft Teams and FaceTime. At first shaky and faltering, these "events" soon developed into a new norm, many aspects of which are likely to remain. Measures first seen as temporary are working well, putting down roots, and sliding into permanence. A new paradigm for wine communication is taking shape, and this is a process that is likely to continue for years to come, streamlining and adapting to suit changing requirements. COVID has advanced the future.

In my own case a roster of in-person events, speaking at wine-themed dinners and so forth, disappeared faster than water down a plughole. Almost overnight, however, commissions flooded in from businesses eager to entertain staff or clients by way of online wine tastings. I now speak to the camera on my laptop and to the array of postage stamp images of attendees in "gallery view." Some organizations proved more adept than others at rapidly adapting to changed circumstances, most notably the London wine club 67 Pall Mall and Vivant, an interactive online platform for wine lovers founded by Silicon Valley entrepreneur and owner of Château de Pommard in Burgundy, Michael Baum.

With its doors shut, 67 turned almost overnight into a virtual club, providing a roster of four or five online tasting presentations a day, every day. There was hardly a wine region, winemaker, or wine expert that did not feature in one of these. A system of sending out sample packs of wine to attendees, prepared under a blanket of inert gas, was quickly developed, with a temperature-sensitive strip to indicate the maximum temperature the wine was exposed to in transit. In the case of the most popular tastings, hundreds tuned in from scores of countries across the globe. This continued for a year until the next logical step—a TV channel—launched in May 2021. Film crews in a dozen countries were commissioned to shoot countless hours of vineyard and winery footage, studios are being built in Bordeaux and Burgundy, in addition to those in London and the newly opened "67" in Singapore, all with the aim of providing twenty-four-hour television to cater for every time zone around the world. Mindful of the fact that wine alone cannot carry a TV show— it tends to "clam up" when placed center stage—food and travel are given equal billing with it.

Vivant, already in the planning phase before COVID struck, champions the interactive nature of its offering, which includes sample tasting packs similar to 67 Pall Mall. Apart from traveling virtually to wine regions to taste the wines and meet the winemakers, attendees can participate by way of simple quizzes that pit them in friendly competition with each other, all facilitated by a team of adept presenters who keep things moving along without any sense of rush. In addition, Vivant makes it its mission to promote organic and biodynamic wines and, with remote working now here to stay, also offers tasting events designed as team building exercises for companies and businesses.

For consumers the advantages of online tastings are legion, as they enable them to "go out" without leaving home: taxis don't have to be booked, nor babysitters, nor do participants need to dress up, and they can participate or not as they wish. Crucially, and most importantly, the attendees can be geographically distant, scattered many miles apart, at a remove that would otherwise preclude their attendance at an in-person event. For this reason alone, the virtual online tasting is here to stay.

"Armchair traveler" engagement with wine will increase, allowing consumers to "visit" wine regions, even as those regions open up again to "real" visitors. Ultimately, nothing beats putting your feet on the ground, but your choice of ground will be greatly facilitated by virtual visits to perhaps three regions before settling on one. Travel will be more considered and as places such as Burgundy and Bordeaux vie to encourage visitors to their expensive new attractions and facilities, what they offer will be improved by competition.

For the wine trade, the travel treadmill of yore may never come back. The trade once traveled ceaselessly: buying, selling, attending trade fairs, visiting vineyards, meeting customers, lapping the globe. Lap after lap. Was it all necessary or had it become an unquestioned grind? As with all businesses it was discovered that much could be done remotely, face-to-face engagement could frequently be dispensed with. In the future, journeys will be more carefully planned, not just to save time and money but to chime with the broader environmental agenda now being set by climate change.

No industry is more aware of the effects of climate change than the wine industry, effects seen painfully in April 2021 when cataclysmic spring frosts hit the vineyards of France, wiping out the nascent crop in some places and severely damaging it in others. The frost was simply the executioner; it was the early arrival of spring, advancing the vines' growth, that pronounced sentence. And that early arrival was the result of climate change, which may yet prove the greatest challenge to face winemakers since phylloxera. Unlike phylloxera, the solution to climate change may not be implemented in time to have any benefit, if the current faltering efforts are anything to go by.

Within the wine trade, many measures being considered, such as lightening the weight of glass bottles to reduce emissions in manufacture and shipping, are little better than sticking plasters. These minor adjustments will only slow down the damage; certain practices must be stopped, not modified. And, as with all difficult change, nothing of note will be done until we are looking down the barrel of a gun. If climate change is to be halted and perhaps even reversed then slowing the runaway vehicle is not good enough; a change of course is needed.

Radical action: if we are truly concerned about the planet, shouldn't we stop shipping wine internationally? Living in a country that produces almost no wine, I am not in favor of such drastic action, but this sort of thing needs to be considered if we want to make an impact. Perhaps wine shipments should be restricted to continent of production? We are constantly being encouraged to shop and eat

A Crazy Thought

Sparkling wine to rival Champagne is now being made in England, and Champagne houses such as Taittinger and Pommery have been quick to acknowledge the potential by buying land there. Perhaps this is a solution of sorts to the challenge of climate change, and, if Champagne can do it, why not Burgundy? At current land prices in Burgundy, a domaine with a few hectares of grand and premier cru vineyards could sell up for tens of millions and probably afford to buy half a county in England. Brexit has put a bureaucratic hurdle in place, but hurdles are for jumping and, coming from France, they will be no strangers to completing mountains of paperwork.

It is more likely that one of the big négociants will dip their toes, as an insurance policy if nothing else. If summers keep getting warmer in Burgundy, the current winemaking model will no longer be fit for purpose, no matter what tweaks are made, but recognition of that will come slowly. Like any industry wrong-footed by change, it is those who are able to re-invent themselves that will survive. And those that start afresh, newcomers unhampered by years of tradition and practices suited to a receding age, will prosper. If a well-resourced Burgundy domaine was to sell up and move to England their descendants could toast their forebears' wisdom for generations to come. Which is a crazy thought, for now.

locally and seasonally. Why not drink locally? Why ship countless millions of liters all over the globe? If we are serious about tackling the problem, then we must do so seriously and abandon cosmetic measures that do little more than salve our consciences.

Over the years, many of my soothsaying efforts have fallen on stony ground, but amid their remains I count one notable success. Dining in a Bordeaux château over a decade ago, I predicted that fashion's finger was about to point to Burgundy and away from Bordeaux as the world's favorite trophy wine. No crystal ball hokum was needed to conjure evidence; it was already happening and was there to see for those who cared to look, though my choice of venue for airing this prediction was questionable. The assembled guests harrumphed but indulged my fantasy as one might a child who blurts gauche opinions.

Fashion is the hidden hand that dictates success for one product and failure for another. Outside the wine world, torn jeans are a hugely successful clothing item, worn by otherwise stylish people unaware of how comical they look. Fashion cannot be accounted for; it can flip in an instant. Fall out of fashion and see how you must cut your prices to sell your wine. Fall into fashion and bemoan the fact that you can never meet demand. It's like the weather—hard to predict, difficult to harness, and impossible to ignore. It must be factored in when future gazing, allied to a think-the-unthinkable mindset. If something is popular now there's a fair chance its opposite will be popular tomorrow. Consider Australian Chardonnay—once rampantly successful, it was dumped when consumers tired of its unctuous charms, so it morphed into wimpish anonymity in response, before finally regaining its mojo, now svelte where it was once bloated. Yet if anybody had predicted its demise during its glory days they would have been howled down, a golden future beckoned, the rough seas were never forecast.

At the high end of the market fashion's influence is immense. As predicted, Burgundy is currently cock o' the walk, with deep-pocketed collectors scrambling to secure the most prized wines for their cellars. Wines made by the late and legendary Henri Jayer now fetch prices that are way beyond mind-boggling. At an auction in 2018—in rough figures—a total of some $35 million was paid for the equivalent of 1,273 bottles that came direct from his own cellar, giving them impeccable provenance. Back-of-an-envelope arithmetic translates those figures into a price of $27,500 per bottle or $180 per teaspoon. Think about it: $180 per teaspoon for fermented grape juice. There was a time when it only took a deep breath and a stiff workout for the credit card to buy a few Jayer bottles; I didn't have to sell the house.

Fashion has more to answer for than high prices for great Burgundies and others of that ilk. More irritating is how it has favored the fortunes of rosé wines of variable quality.

I've had it with rosé. The love affair, never better than lukewarm, has dribbled to a close. Only bafflement, that so many are still enchanted by its limp glow and faux sophistication, remains. I'm tired of feigning polite interest in pastel-colored and pastel-flavored wine. The prices paid for the high-end ones are crazy—think what wonderful red or white might be got for the same price. I once dutifully plodded through a dozen samples, searching for some point of distinction, a hook to hang an opinion on. None was found, unless you count a dreary cosmetic trickle running through them all that settled uneasily in the throat. Rampant and perplexing success has attended rosé for too long. The pink bandwagon is currently burdened with every manner of iteration, too many of which are redolent of cheap perfume.

In a weird paradox, only explained by fickle fashion, there is a baffling price race to the top, as producers compete for the dubious accolade of "world's most expensive rosé." Why this price gouging is not met with scorn and laughed into shameful retreat says much for the marketing people who have spun it into some sort of achievement. In this they have been helped by commentators turned cheerleaders, happy to suspend the critical judgment they apply to other wines. All who trumpet the rise of ludicrously priced rosé should be hiding their shame at promulgating such nonsense, championing it as some sort of achievement, when it is anything but.

Whispering Angel, one of the pack leaders, is even available in four-bottle jeroboams. What other ordinary wine would people buy a jeroboam of? How many are even available in this monster measure? What madness has possessed otherwise sane people

Clean Wine

Being definition-less is no impediment to this new style that comes with its own halo of rectitude. It's even more evocative than "natural." This stuff must be good—it's clean!—and isn't clean the opposite of dirty: a clean sheet, a clean bill of health, a clean sweep? Clean is a "good" word, whereas "dirty" belongs in the naughty corner. This is "safe space" wine, unchallenging, the liquid itself being of no consequence. In terms of nomenclature, "clean" and "natural" are in the same category, naming themselves with a feel-good word to gain some reflected glory from the automatic approbation granted to the word itself.

But natural wines have something to say for themselves; you can push against them, like them or loathe them, have an opinion about them. All I can muster about clean wine is a weary sigh. Supposedly the term references the production process that eschews the use of chemicals, while piling on the vacuous blah-blah. If they take off, what next? Perhaps future wine styles will be defined by ever more mushy names. Are we going to see Divine wines? Inspirational? Pristine? Mannerly? Immaculate? Courteous? We can only hope not. "Clean wine" is a meaningless, empty term. Move on.

to do so? Even Champagne is rarely seen in this format. If you are going to splash out in this fashion, for heaven's sake choose something that won't make you cringe with embarrassment in a few years' time.

Perhaps Champagne is to blame for the pricing, for its pink iteration is always inexplicably more expensive than the regular version. Even though I love the stuff, I object to the price premium they manage to get away with. Are there any rosés that can tempt me out of curmudgeons' corner? Sancerre rosé, made from Pinot Noir, can be delicious on a summer's day outdoors, bereft of the saccharine slick that corrupts so many from warmer climes and lesser grapes.

Have we entered a new puritan age, predicated upon "thou shalt not" and peopled by myriad fun-free strictures? Current advice on safe drinking limits suggests we have. Much of it relies on diktat and proscription to jolt people into line. These are then followed by the woolliest advice, dispensed by doctors and scientists whose work and research requires rigor and exactitude. Strangely, they appear unable to bring those qualities to bear on their advice. Vacuous guidelines such as "a glass of wine" or "a measure of spirits" are bandied about, with no thought for how meaningless they are. Unless we know the size of the glass or measure and the alcoholic strength of the wine or spirit, we know almost nothing. This might sound like pedantic hair-splitting, but if the issue is to be dealt with seriously, then official pronouncements should be clear and precise, and not in need of interpretation that makes people less likely to follow them.

If drinking advice is to be truly useful in the future, it should rely on easily implemented precepts that encourage a move toward moderation, without recourse to joyless pulpit proclamations likely to be ignored by the people they are meant to help. It should consist of practical tips derived from real life rather than highly structured research. Three simple precepts, when combined, can nudge behavior in the right direction, especially for those concerned by an increase in their consumption as a result of more time spent at home because of COVID restrictions.

First, develop a habit of pouring less wine at a time and then leave the bottle out of reach so that you have to leave your seat for a refill. And beware of currently fashionable large glasses—a reasonable pour looks little larger than a wine splash in some of them. Second, actively seek out lower alcohol styles, not de-alcoholized wines, which are maimed by manipulation, their deficiencies masked by right-on approbation rather than any critical judgment. Hobbled into harmony with the current zeitgeist, they are having a moment

The One Bottle Parable

Seeking out lower-alcohol wines is the best way to moderate intake, but it's easier said than done. Too often the percentage is given in tiny figures printed in an ink that barely contrasts with the paper of the label, mauve on purple for instance. Thus it is easy to miss, yet paying attention can yield impressive rewards.

Take an apocryphal couple who have shared a standard bottle of wine with their Friday dinner for the last thirty years. If asked, they can say, hand on heart, that their wine consumption has not increased in that time. It hasn't, but if the bottle thirty years ago was 12 percent alcohol and today's is 15, then their alcohol consumption has increased by 25 percent. Figures of 12 and 15 percent might stretch the envelope a little but not much. By switching back they can immediately reduce their alcohol consumption by 20 percent, with no reduction in enjoyment, perhaps even an increase, by way of a lighter style of wine with less whack on the palate. It pays to examine those tiny little numbers at the bottom of the label.

now but, dressing hope as prediction, I'm saying their time will pass. Styles that are naturally lower include: Hunter Valley Semillon, Mosel Riesling, and many Loire Valley wines. This is the most painless way to reduce consumption. Third, don't feel any need to finish the bottle; it will keep overnight. Indeed, it will probably keep for a few days. An unfinished bottle of wine lasts longer than commonly thought. We need to banish any residual belief that a bottle needs finishing on the day it is opened. I frequently stretch a bottle over three or four days to see how it develops, pushing the boundaries and only occasionally tripping myself up. The bonus is that drinking a wine over a couple of days reveals it completely, hidden nuances are revealed, unseen weaknesses too. The full picture emerges. Finally, have a glass of water on hand to quench thirst, so as not to do so with wine.

I am not ignoring the fact that alcohol can be an addictive, damaging substance, but if pronouncements on safe levels of drinking are to be effective then they need to be couched as above. Grim epistles outlining a difficult-to-achieve ideal have little value.

Hindsight. We revel in its clarity while wishing foresight was similarly lucid. Everything in the rearview mirror is clear while the road ahead is murky. Humans don't like uncertainty and attempt to throw parameters around the future with carefully laid plans. It's like using twigs to dam a river. In 2020 and 2021 COVID burnt the plans. Doubt and uncertainty came center stage, dogging every step, infecting every thought, plan, opinion, sentiment, emotion. Uncertainty was never writ so large. Plans were made and unmade. Notwithstanding that, it is impossible not to take a glance into the crystal ball in the hope of spotting what's coming down the track. Soothsaying never loses its appeal.

If climate change proves irreversible, vineyards will migrate, painful and all as that will be. Wine production is already drifting north in Europe, with vines now planted in places such as Poland, Belgium, and Denmark. More will follow. The great imponderable is China—both as producer and consumer. Whatever path China takes it will be hugely influential, as witness the major ripples she caused in late 2020 when slapping massive tariffs on Australian wine imports. China wields huge power and if she ever chooses to export wine in earnest, the wine world will be immediately reshaped.

Perhaps it already has been reshaped. The knowledge burden has made us over-prissy about wine. We must become less rule-bound; it is time to lighten up. High-end wines need cosseting, but too much sanctimony has filtered down to the simple stuff. The simple wines need no ceremony; serve them from kegs, put them in bag in box, or even can them. Make them clean and fresh, pure and

The Year 2222

It is two hundred years hence and the wine world has changed. Burgundy is now made from Syrah and in a style said to resemble the Barossa Valley Shiraz of the late twentieth century. Genuine Champagne is made in England, the rights to the name being sold after the Franco-German war left the vineyards polluted beyond rescue. Poland, once renowned only for vodka, is now the go-to source for the best Bordeaux blends. Some liken them to the great clarets of old, of which a lingering memory remains. Ireland, similarly renowned for whiskey and dark beer, is famed for Blackwater Valley Riesling.

Bordeaux itself is now a workhorse region, arid and irrigated, churning out strapping red wines whose only attribute is the heft they add to the rather weedy North Cape wines from Norway. Some bottles of Cabernet Sauvignon remain from California's last viable harvest fifty years ago. Thereafter, the few remaining wineries were abandoned because of endless wildfires. Canada is the new France, world leader in a host of styles and one that is uniquely her own: Cabrah, a late twenty-first century cross between Cabernet Franc and Syrah. Tasmania is the only remaining wine region in Australia. And if you want to be really on the money, Danish Gamay is the wine du jour.

juicy. Above all, endow them with some character, then pour and enjoy. Be practical, not prissy.

I will remain wedded to the traditional bottle, treated with due reverence, cellared on its side, and opened with appropriate ceremony to be enjoyed with a diminishing band of like-minded souls. But if packaging alternatives and a more free and easy attitude ensure wine's continued popularity into the future, then I am all for it. Wine is being challenged like never before by alternatives such as limited-release whiskies, craft beers, and, principally, ever more outlandish cocktails made with small-batch gins of high price and dubious merit.

Within the wine world the easy categories of old are no longer clearly defined. White wine is now orange, red is black, rosé is everywhere. The classics are being pushed to the margins. Or are they? I was leaning, reluctantly, toward this opinion when a bottle of Château Léoville Barton 2005 brought me up short. It was like a pure, clear voice in a sea of babble: balanced, harmonious, and elegant. Class will always out.

We as consumers should celebrate the classic and then seek out the characterful. Dare to stray from the middle of the road in this quest, chase up the side alleys, otherwise the middle will deepen into a rut from which escape will be impossible. Wine is an endlessly fascinating subject, sometimes frustrating, more often rewarding and, crucially, never ending, there is always something else to be discovered. But it must be discovered. It is true that more good wine than ever before is being made but the good stuff needs a market to survive.

Wine has enjoyed a recent golden age which, if a single marker can be pointed to, opened forty years ago with the excellent 1982 vintage in Bordeaux, wine's capital. That age is now variously threatened, including from within by ludicrous prices, both high and low. If the stuff at the bottom end has no character and the expensive stuff is out of reach then the next generation of consumers will look elsewhere for palate pleasure. Wine should be a big church, accommodating of diversity, with room for everybody. Is the golden age drawing to a close? It will be mid-century before we know.

INDEX

A

Albariño, 18, 179
Alcock, John, 161
Aldo Conterno, 77
Alicante Bouschet, 75–76
Aligoté, 156, 177
Allergen Index, 152
Almaviva, 22
Alsace, 166–168
Amontillado, 55
Auslese, 169
Australia, 4, 79–81, 85–91, 93–94

B

Barbaresco, 23
Barolo, 9, 23, 77, 110–111, 157

Barossa Shiraz, 170, 190
Barret, Matthieu, 40
Bass Phillip, 86
Beaujolais, 120–122
Beerenauslese, 169
Black Tower, 25
Blackwater Valley Riesling, 190
Blaufränkisch, 133
Blue Nun, 25
Blush Zinfandel, 42
Boal, 59–60
Bollinger, 29, 33
Bordeaux, 4–5, 19, 21–22, 65, 68, 100, 118, 181, 184, 191
Bostwick, William, 63
Bride Valley, 43

Brouilly, 122
Brown, Arthur, 161
Bual, 58
Burgundy, 7, 12–13, 15, 19, 21, 24, 76, 100, 107, 113, 118, 181, 183–184

C
Cabernet, 94
Cabernet Franc, 65, 118, 190
Cabernet Sauvignon, 65–66, 68, 70, 101, 106, 111, 157, 190
Carrodus, Bailey, 88–89
Cava, 41–43
Chablis, 156
Champagne, 27–41, 43–45, 53, 64, 131, 136, 146, 159–160, 183, 186–187
Chardonnay, 1–3, 36, 38, 45, 80–81, 184
Charmes-Chambertin, 77
Chassagne-Montrachet, 7, 121
Château Chasse-Spleen, 1–4
Château de Pommard, 73, 180
Château du Nozet, 116
Château d'Yquem, 161–162
Château Haut-Brion, 105
Château Lafite, 133, 146
Château Léoville Barton, 191
Châteauneuf-du-Pape, 8, 107
Château Pavie, 107
Château Rayas, 108
cheese, 156–157
Chénas, 122

Chenin Blanc, 119, 156
Chianti, 110–111
China, 91, 113, 189
Chinon, 118
Chiroubles, 122
Clare Valley, 92–93
Clos des Papes, 8, 158
Clos Sainte Hune, 167
Cloudy Bay Sauvignon Blanc, 94–95, 140
Conch y Toro, 22
Conklin, Ted, 66
Coppola, Francis Ford, 71
Coravin, 138–139
corkscrews, 130–132
Corton-Charlemagne, 140
Coteaux Champenois, 40–41
Côte de Brouilly, 122
Côte d'Or, 12, 16, 38, 72, 76, 121
Côte Rôtie, 8
Crémant, 39–40
Crémant de Alsace, 39
Crémant de Bordeaux, 39
Crémant de Bourgogne, 39
Crémant de Loire, 39
crubeens, 155

D
decanter, 137
Domaine de la Pousse d'Or, 113
Domaine de la Romanée-Conti, 73–74
Domaine Dugat-Py, 77

Domaine Dujac, 73
Domaine du Vieux Télégraphe, 8
Domaine Huet, 8–9
Domaine Huet Vouvray, 140
Douro, 15–16, 48
Drop Stop, 143

E
Eiswein, 169
Emilia-Romagna, 42
Eszencia, 165, 172–173

F
Finger Lakes, 63–65
Finger Lakes Riesling, 9
Fino, 55–56, 60, 145–146, 156
Fleurie, 122
Franciacorta, 42
Frank, Konstantin, 63–64

G
Gago, Peter, 84
Gaja, Angelo, 110
Galicia, 17–18
Gallo, Gina, 73
Gamay, 121, 190
Gevrey-Chambertin, 7, 86
Gewurztraminer, 167
glasses, 132–138
Glera, 41
Godello, 179
Gramona, 43
Grande Cuvée, 34

Grenache, 70, 108
Grillo, 57
Guiraud, 163
Gusborne, 43

H
Hambledon, 43
Hill of Grace Shiraz, 91–92
Hungary, 164–165
Hunter Riesling, 82
Hunter Valley, 80–81
Hunter Valley Semillon, 9, 82, 188
Hyde de Villaine, 73

I
Ireland, 91, 161

J
Jayer, Henri, 86, 184
Jefferson, Thomas, 77–78
Johnson, Hugh, 179
Jones, Phillip, 86
Judd, Kevin, 95
Juliénas, 122

K
Kabinett, 169
Krug, 33–34

L
labrusca, 64
La Mancha, 25
Lambrusco, 42

Laurent-Perrier, 33
Lehmann, Peter, 90–91
Leroux, Benjamin, 26
Liebfraumilch, 25, 123
Liger-Belair, Thibault, 12
Loire Valley, 116, 118–120, 188
Lovedale, 82

M
Macabeo, 42
Madeira, 47, 58–61, 171
Malbec, 142, 157
Malmsey, 58
Man O' War Dreadnought Syrah, 96–97
Manzanilla, 55, 60, 145
Marlborough Sauvignon Blanc, 95
Marqués de Riscal, 9
Marsala, 57
Mascarello e Figlio, Giuseppe, 9
Mega Purple, 19–20
Merlot, 70
Meursault, 13, 156
Miles, 69–70
Moët et Chandon, 33
Mondavi, Robert, 71–72
Montebello Cabernet, 69
Montrachet, 143
Morgon, 122
Mosel Riesling, 90, 122–125, 168–169, 188
Mosel Valley, 64
Moulin-à-Vent, 122
Mouton Rothschild, 22

Müller, Egon, 124, 168
Müller-Thurgau, 95
Mumm, 131
Muscadet, 18, 117

N
Napa Valley Cabernet Sauvignon, 68
Nebbiolo, 23, 111
New Zealand, 94–96, 164
Niebaum, Gustave, 71
Nuits-Saint-Georges, 12
Nyetimber, 43

O
Oloroso, 55–56
Opus One, 22
Oregon, 78–79

P
Palo Cortado, 55–56
Palomino, 55
Pantelleria, 17
Parellada, 42
Paumanok, 66
Pedro Ximénez, 56
Penfolds Grange, 22, 83
Pewsey Vale Contours, 92
phylloxera, 72–74, 90, 182
Pickett, Rex, 69
Piedmont, 22–23
Pineau de la Loire, 119
Pink Port, 50
Pinot Blanc, 167

Pinot Gris, 156

Pinot Meunier, 36

Pinot Noir, 36, 44–45, 76, 86, 95–96, 101

Pol Roger, 33, 53

Port, 5–7, 47, 49–52

    Warre's Vintage Port, 1

Pouilly-Fumé, 116

Prohibition, 75–76

Prosecco, 41

R

Ramonet, Jean-Claude, 143

Ravenswood Cooke Zinfandel, 67

Recaredo, 43

Régnié, 122

Reims, 38–39

Reisinger, Andrew, 63

Rhône Valley, 8

Rías Baixas, 10

Ribeira Sacra, 10

Ribera del Duero, 105, 109

Ridge, 69

Riesling, 8–9, 39, 82, 90, 92, 122–125, 140, 166–168

Rieussec, 163

Rioja, 9, 109, 131, 157

Riverina, 20

Robinson, Jancis, 136

Roederer, 33

rosé, 185

Rosemount Chardonnay, 1–3

Rousseau, Charles, 86

Rubired, 19

Ruinart, 33

Rutherglen Muscat, 170

Ryan, Phil, 82

S

Saint-Amour, 122

Sancerre, 116

Sassicaia, 110

Sauternes, 81, 156, 162, 164, 166

Sauvignon Blanc, 94, 107–108, 164

Scharzhofberger, 124

Schubert, Max, 83–85

Semillon, 81, 94, 97

Seña, 22

Sercial, 58

Shakespeare, William, 52

Sherry, 8, 17, 47–48, 52–61, 145–146, 156, 173

Shiraz, 80, 90–91

South Africa, 107–108, 119, 171–172

sparkling wine. See Champagne; Crémant; Riesling

Spätlese Trocken, 169

Spurrier, Steven, 21

Standalone Cabernet, 96

stemless wineglasses, 135

Suduirant, 163

Sugrue, Dermot, 10

surf'n'turf, 147

Syrah, 9, 40, 95–96, 101, 108, 190

T

tattoos, 148

Tawny Port, 50–51

Tempranillo, 108–109

Tignanello, 110

Tio Pepe, 53

Tokaji, 9, 164–165

Trockenbeerenauslese, 169–170

Trotter, Charlie, 149

Tuthills Lane Block, 67

V

Vega Sicilia, 9, 105

Veltliner, Grüner, 133, 150

Verdejo, 179

Verdelho, 58

Vidigal Wines, 26

Villaine, Aubert de, 73

Vin de Constance, 171–172

Viognier, 78

Virginia, 77–78

Viura, 179

Volstead Act, 75

Vosne-Romanée, 7

Vouvray, 8–9, 119–120, 140

W

"waiter's friend," 131–132

Whispering Angel, 185–186

Wild Earth Pinot Noir, 95

wineglasses, 132–138

Wiston, 10, 43

X

Xarel-lo, 42

Y

Yarra Pinot, 170

Yarra Yering, 88–89

Z

Zinfandels, 67–68, 115

Zweigelt, 133